Science Uncovered

AQA
Chemistry
for GCSE

Moira Sheehan
Martin Stirrup

WITHDRAWN

Series Editor: **Keith Hirst**

www.heinemann.co.uk
✓ Free online support
✓ Useful weblinks
✓ 24 hour online ordering

01865 888058

D0334265

* 000263791 *

Heinemann Educational Publishers
Halley Court, Jordan Hill, Oxford OX2 8EJ
Part of Harcourt Education

Heinemann is the registered trademark of Harcourt Education Limited

First published 2006

10 09 08 07 06
10 9 8 7 6 5 4 3 2 1

10-digit ISBN: 0 435 586068
13-digit ISBN: 978 0 435 586065

Copyright notice
All rights reserved. No part of this publication may be reproduced in any form or by any means (including photocopying or storing it in any medium by electronic means and whether or not transiently or incidentally to some other use of this publication) without the written permission of the copyright owner, except in accordance with the provisions of the Copyright, Designs and Patents Act 1988 or under the terms of a licence issued by the Copyright Licensing Agency, 90 Tottenham Court Road, London W1T 4LP. Applications for the copyright owner's written permission should be addressed to the publisher.

Edited by Patrick Bonham, Linda Moore and Anne Trevillion
Designed by Lorraine Inglis
Typeset by Ken Vail Graphic Design

Original illustrations © Harcourt Education Limited, 2006

Illustrated by Beehive Illustration (Martin Sanders, Mark Turner), Nick Hawken, NB Illustration (Ben Hasler, Ruth Thomlevold), Sylvie Poggio Artists Agency (Rory Walker).

Printed by CPI Bath

Cover photo: © Science Photo Library

Index compiled by Indexing Specialists (UK) Ltd

Picture research by Zooid Pictures Ltd

Acknowledgements
The authors and publisher would like to thank the following individuals and organisations for permission to reproduce photographs:

Keith/Custom Medical Stock Photo/Science Photo Library p iv R; Peter Gould/Harcourt Education p iv L; Tim Ayers/Alamy p 2 T; Getty Images/PhotoDisc/ p 2 BR; Carol Dixon/Alamy p 2 BL; Jason Hawkes/Corbis UK Ltd. p 3 R; Getty Images/PhotoDisc/ p 3 L; Leslie Garland Picture Library/Alamy p 4; Chris Howes/Wild Places Photography/Alamy p 6 T; Josè Manuel Sanchis Calvete/Corbis UK Ltd. p 6 B; Getty Images/PhotoDisc p 8 TR; Gillian Darley; Edifice/Corbis UK Ltd. p 8; Alistair Berg/Alamy p 8 TL; Paul Hardy/Corbis UK Ltd. p 9; Nick Gregory/Alamy p 10 T; Martin Stirrup p 10 B; Chris Henderson/Construction Photography p 11 T; Corbis p 11 B; Index Stock Imagery/Photolibrary.com pp 12, 93 B; Lee Pengelly/Alamy p 13 T; Biophoto Associates/Science Photo Library p 14 TR; Vaughan Fleming/Science Photo Library p 14 TL; Charles Bowman/Photolibrary.com p 14 B; Corbis p 15 T; Marie-Louise Avery/Alamy p 15 B; Getty Images/Brand X Pictures/ p 16 T; Jesper Jensen/Alamy p 16 B; Paul Thompson; Eye Ubiquitous/Corbis UK Ltd. p 17; Dan Sinclair/Zooid Pictures p 19; Tony Waltham/Robert Harding Picture Library Ltd/Photolibrary.com p 20 B; Charles D. Winters/Science Photo Library p 20 T; Michael Barnett/Science Photo Library p 21; Getty Images/PhotoDisc p 22 T; Harmon Maurice/Images.Com/Photolibrary.com p 22 M; James L. Amos/Corbis UK Ltd. p 22 B; Science Photo Library/Science Photo Library p 23 T; Paul Seheult/Eye Ubiquitous/Corbis UK Ltd. p 23 B; NASA/Photolibrary.com p 24 B; Joel Stettenheim/Corbis UK Ltd. p 24 T; Alan Novelli/Alamy p 26 TL; Peter Turnley/Corbis UK Ltd. p 26 TR; Reuters/Corbis UK Ltd. p 26 B; Corbis p 30; POPPERFOTO/Alamy p 32; Paul Glendell/Alamy p 33; Pablo Corral Vega/Corbis UK Ltd. p 34; Getty Images/PhotoDisc p 35; Martin Bond/Science Photo Library p 36; Botanica/Index Stock Imagery/Photolibrary.com p 42 B; Cordelia Molloy/Science Photo Library p 42 T; Visual Arts Library (London)/Alamy p 46; NOGUES ALAIN SYGMA/Corbis UK Ltd. p 48; AJ Photo/Science Photo Library p 50; Zooid Pictures p 52; TH Foto-Werbung/Photolibrary.com p 54 T; Sheffer Visual Israel/Photolibrary.com p 54 B; Gerry Johansson/Bildhuset Ab/Photolibrary.com p 55; IPS Photo Index/Ips Co Ltd/Photolibrary.com p 56 L; Frank Wieder/Photolibrary.com p 56 R; Bryan & Cherry Alexander Photography/Alamy p 57; David R. Frazier Photolibrary p Inc/Alamy p 57; Getty Images/PhotoDisc p 58; Garry Gay/Alamy p 58; Getty Images/PhotoDisc p 60; Michael Carter Photography/Photolibrary.com p 61; Peter Sapper/Photolibrary.com p 62; Tomas del Amo/Phototake Inc/Photolibrary.com p 63; Jon Arnold Images/Photolibrary.com p 65; Getty Images/PhotoDisc p 66; AP Photo/Eric Skitzi/Empics p 67 T; Corbis UK Ltd. p 67 B; Corbis p 69; Gali Danielle/Jon Arnold Images/Photolibrary.com p 70; Corbis UK Ltd. p 71; Illustrated London News p 73; Werner Otto/Alamy p 82 T; Sheila Terry/Science Photo Library p 82 B; Sheila Terry/Science Photo Library p 83 T; Tek Image/Science Photo Library p 83 B; Mary Evans Picture Library p 87; Andrew Lambert Photography/Science Photo Library p 90 L; Roger Scruton p 90 R; Elio Ciol/Corbis UK Ltd. p 94; Peter Gould p 95; Boart Longyear Europe p 96 T; Paul Whitehill/Science Photo Library p 96; KPT Power Photos p 96; Getty Images/1053 p 96; Getty Images/5162 p 96 B; John Giles/PA/Empics p 98; James King-Holmes/Science Photo Library p 100 T; Science & Society Picture Library p 100 B; sciencephoto/Alamy p 101 TL; Phil Schermeister/Corbis UK Ltd. p 101 TR; Digital Vision/ p 101 B; Trevor Clifford/Harcourt Education p 104; Charles D. Winters/Science Photo Library p 106; Peter Gould p 107 T; Caroline/Zefa/Corbis UK Ltd. p 107 B; Peter Gould p 108; Peter Gould p 109; Andrew Lambert Photography p 110 L; Science Photo Library p 110 R; Corbis/ p 111; Science Photo Library p 112 T; Peter Gould p 112 M; Peter Gould p 112 B; Corbis/ p 114 T; Science Photo Library p 114 B; Zooid Pictures p 117; Chinch Gryniewicz; Ecoscene/Corbis UK Ltd. p 118 T; Stefan Rousseau/Pa Pool/Reuters/Corbis UK Ltd. p 118 M; AP Photo/Claude Petit/Empics p 118 B; Sipa Press/Rex Features p 119 L; Sipa Press/Rex Features p 119 R; Unknown p 120; Tudor Photography/Harcourt Education p 122; Peter Gould p 123; Peter Gould p 124; Document General Motors/Reuter R Sygma/Corbis UK Ltd. p 127 T; Zooid Pictures p 127 B; Braun p 128; NASA p 129; Stacey/Fox Photos/Hulton Archive/Getty Images p 131 L; PhotoDisc/ p 131 M; Sony Ericsson UK & Ireland p 131 R; Peter Gould p 132 T; Peter Gould p 132 BR; Roger Scruton p 132 BL; Matt Olsen/iStockphoto p 133 TL; Creatas/ p 133 TR; Peter Gould p 133 B; Trevor Clifford/Harcourt Education p 134 T; Leslie Garland Picture Library/Alamy p 134 M; Getty Images/AA012667/ p 134 B; Jason Hawkes/Corbis UK Ltd. p 135; Martyn F. Chillmaid p 136 T; Peter Gould p 136 B; Hulton-Deutsch Collection/Corbis UK Ltd. p 140 T; Zooid Pictures p 140 B; Digital Vision/ p 142 T; Martyn F. Chillmaid/Science Photo Library p 142 B; Geoscience Features Picture Library p 143 L; Geoscience Features Picture Library p 143 R; Peter Gould p 145 T; Martin Bond/Science Photo Library p 145 B; David Muench/Corbis UK Ltd. p 146 T; Geophotos p 146 B; Holt Studios International p 148; Peter Gould p 149 T; Robert Brook/Science Photo Library p 149 B; Peter Dean/Agriculture.com p 152; Getty Images/72133 p 162; Bettmann/Corbis UK Ltd. p 165; Martyn F. Chillmaid/Science Photo Library p 168 T; Martyn Chillmaid/Photolibrary p 168 TM; Andrew Lambert Photography/Science Photo Library p 168 M; Theodore Gray/RGB Research Limited p 168 BM; Theodore Gray/RGB Research Limited p 168 B; sciencephotos/Alamy p 169 B; Charles D. Winters/Science Photo Library p 169 T; Martyn F. Chillmaid p 170; sciencephotos/Alamy p 171; DACS 2006/AKG-Images p 172 L; sciencephotos/Alamy p 172 T; sciencephotos/Alamy p 172 TM; sciencephotos/Alamy p 172 BM; Martyn F. Chillmaid p 172 BM; sciencephotos/Alamy p 172 B; Used with permission. From Chemistry Comes Alive! video collection/Journal of Chemical Education Software p 174; Theodore Gray/RGB Research Limited p 176 a; Klaus Guldbrandsen/Science Photo Library p 176 b; GeoScience Features Picture Libr/Geoscience Features Picture Library p 176 c; Charles D. Winters/Science Photo Library p 176 d; Charles D. Winters/Science Photo Library p 176 e; Geoscience Features Picture Library p 176 f; Charles D. Winters/Science Photo Library p 176 g; Lester V. Bergman/Corbis UK Ltd. p 176 h; E.r.degginger/Science Photo Library p 176 i; Hans-Frieder & Astrid Michler./Science Photo Library p 176 j; V&A Images/Alamy p 177 T; J & L Weber/Still Pictures p 177 M; Ron Hohenhaus/iStockphoto p 177 B; Michael S. Yamashita/Corbis UK Ltd. p 178 T; Jet Propulsion Laboratory (NASA-JPL)/NASA p 178 B; Ian West/Photolibrary p 180 L; Niall Benvie/Corbis UK Ltd. p 180 R; Bob Krist/Corbis UK Ltd. p 181 T; Trip/Alamy p 181 B; Zooid Pictures p 182; Martyn F. Chillmaid p 184; Soqui Ted Sygma/Corbis UK Ltd. p 185; Martyn Chillmaid/Photolibrary p 186 T; Michael Booth/Alamy p 186 M; Martyn F. Chillmaid p 186 B; Andrew Lambert Photography/Science Photo Library p 189 T; Photolibrary p 189 B; Martyn F. Chillmaid/Science Photo Library p 190; Reuters/Corbis UK Ltd. p 192; Getty Images/6025/ p 193; Alison Wright/Corbis UK Ltd. p 196; Peter Hulme; Ecoscene/Corbis UK Ltd. p 198 T; Chris Lisle/Corbis UK Ltd. p 198 B; Potters for Peace p 199 T; Paul Almasy/Corbis UK Ltd. p 199 B; Dan Sinclair/Zooid Pictures p 200 L; Dan Sinclair/Zooid Pictures p 200 R; Sheila Terry/Science Photo Library p 201; Jason Reed/Reuters/Corbis UK Ltd. p 204; United States Civil Air Patrol p 205; scott elliott/iStockphoto p 206 T; Fire Testing Technology Limited p 206 B; Phil Degginger/Alamy p 207 T; Dr. Richard Chapleau p 207 B; Getty Images/OS01005/ p 208 L; Getty Images/OS01006/ p 208 R; NASA p 212; Aero Graphics, Inc./Corbis UK Ltd. p 213; Bettmann/Corbis UK Ltd. p 214; BMW p 215 T; General Motors Europe p 215 B; geogphotos/Alamy p 216; Sally A. Morgan; Ecoscene/Corbis UK Ltd. p 217; Getty Images/34191/ p 218 T; Phototake Inc/Photolibrary p 218 B; Dept. Of Physics, Imperial College/Science Photo Library p 219; Jose Luis Pelaez, Inc./Corbis UK Ltd. p 220; Lester V. Bergman/Corbis UK Ltd. p 221 Ta; Martyn F. Chillmaid p 221 Tb; Andrew Lambert Photography/Science Photo Library p 221 Tc; sciencephotos/Alamy p 221 Td; Geoscience Features Picture Library p 221 Te; Martyn F. Chillmaid p 221 Ba; Andrew Lambert Photography/Science Photo Library p 221 Bb; Andrew Lambert Photography/Science Photo Library p 221 Bc; David Fleetham/Photolibrary p 222 T; Andrew Lambert Photography/Science Photo Library p 222 M; Martyn F. Chillmaid/Science Photo Library p 222 B; Jerry Mason/Science Photo Library p 223 T; Used with permission. From Chemistry Comes Alive! video collection/Journal of Chemical Education Software p 223 B; Jerry Mason/Science Photo Library p 224; Martyn F. Chillmaid p 226 T; Sherwood Scientific Limited p 226 B; Bettmann/Corbis UK Ltd. p 228 TL; Getty Images/72072/ p 228 TR; Attila Balaz/iStockphoto p 228 M; Charles E. Rotkin/Corbis UK Ltd. p 228 B; Jerry Mason/Science Photo Library p 231.

The authors and publisher would like to thank the following individuals and organisations for permission to reproduce copyright material:

British Water, p 200. All Crown Copyright material is reproduced with the permission of HMSO and the Queen's Printer for Scotland.

Every effort has been made to contact copyright holders of material reproduced in this book. Any omissions will be rectified in subsequent printings if notice is given to the publishers.

Tel: 01865 888058   www.heinemann.co.uk

How to use this book

This book has been designed to cover the new AQA GCSE Science curriculum in an exciting and engaging manner and is divided into four units: C1a, C1b, C2 and C3.

The book starts with two double page spreads focusing on 'How science works', which show you how scientists investigate scientific issues, including those in our everyday lives.

Each unit in the book is broken down into separate sections, for example unit C1a consists of sections 1 and 2. Each section is introduced by a double page introductory spread which raises questions about what is covered in the section, acts as an introduction to the section, and includes a box encouraging you to think about what you are going to learn in the section.

The introductory spread is followed by double page content spreads which cover what you need to learn, but which also cover 'How science works' and the procedures you need to be familiar with to enable you to produce your internally assessed work: the Practical Skills Assessment and the Investigative Skills Assignment.

Each section includes at least one 'ideas, evidence and issues' spread which either focuses on interpreting data and evidence or on evaluating the role of science in society and the issues that affect all our lives.

Throughout the content pages there are in-text questions to test your understanding of what you have just learnt and to further your appreciation of how science can be used and what the issues are surrounding the development of science and technology.

At the end of each unit there are two double page spreads of questions to test your knowledge and understanding of the module. They will also prepare you for the kinds of questions you will meet in exams.

The words displayed in bold in the text also appear in the glossary at the end of the book, together with a definition.

Contents

C3

How science works

▲ Not all scientists work in laboratories.

Why study science?

Even if you are not going to become a scientist, it is important that you know how scientists work. Science affects almost every aspect of our lives – sometimes in very obvious ways, like the development of new technology or new drugs that can be used to treat us, and sometimes in less obvious ways, like additives in our food that we may be unaware of.

We need to know what scientists are up to, so that they cannot 'pull the wool over our eyes' or 'blind us with science'. We need to be able to understand the way in which scientific experiments are carried out and the way in which information is collected. We need to be able to tell the difference between facts and opinions and to judge whether the information, or the people providing it, are biased in any way. Then we can make our own judgements based on an understanding of the facts.

Scientists frequently use a number of technical terms, and often these have meanings that are slightly different from the meanings in everyday speech. These terms have been printed in **bold** and are explained in the text and in the Glossary.

General principles

Many scientific investigations start with **observations**. These need to be made carefully and should be unbiased. They are often the basis for investigations or classification of things.

One of the skills that you need to learn is the ability to distinguish between scientific fact and opinion.

Local residents noticed that large numbers of fish seemed to be dying in the river. The river flows past several factories but then flows through farmland. They suspected a detergent factory was the cause of the pollution.

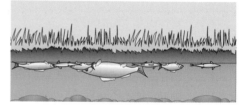

▲ Observing the fish dying in this river started a local investigation.

▲ This factory discharges waste into the river.

River pollution blamed on detergents factory

Local residents have become alarmed at the number of fish dying in the River Sabrina, and have blamed the detergents factory of Lees & Co. Local farmer Fred Guest, who has lived by the river for 30 years, said, 'I've never seen anything like it. I know it's caused by Lees because they use chemicals that contain cyanide, and I've seen them discharging waste into the river.'

A spokesperson at Lees & Co. said, 'It's nothing to do with us. The cause of the problem is all the nitrates that local farmers are putting on the land in fertilisers. They get washed down into the river. There was never any problem until recently – they used to use natural manure.'

Question

a Are the statements made by
(i) Farmer Fred Guest, and
(ii) the spokesperson for Lees & Co. giving facts or opinions? Give a reason for your answer.

As a result of protests, a team of scientists was called in to carry out an investigation. They decided to sample the water in the river at different places.

Question

b What do you think that the scientists should be measuring in the water?

Designing an investigation

When you are designing an investigation you need to plan carefully to make sure that you are measuring the correct **variables** at suitable **intervals** (length of time between measurements) and over a suitable **range**. You need to be ready to measure values between the upper and lower ones that will occur, for example daytime temperature.

The **independent variable** is the one that you deliberately alter. The **dependent variable** is the one that changes as a result of this. Other variables may also affect the outcome, and these need to be controlled or monitored. These are called **control variables**.

You also need to think about whether you need to repeat any of the measurements to improve **reliability**. A value becomes more reliable the more times it is measured, for example taking a reading five times and working out an average.

It is important your results and conclusions are **valid**. Valid results are ones that answer the original question. A valid result can be matched by other scientists following your method. They should get the same result, which validates your result.

The scientists sampled the water at four different points.

The scientists also needed to choose their measuring instruments carefully. In any investigation, it is important to make measurements using the most suitable instruments. Different instruments measure to different levels of **precision**. A tape measure measures to the nearest 1 centimetre and a ruler to the nearest 1 millimetre. You also need to use instruments effectively to get **accurate** results. An accurate reading is one that is nearest to the true result.

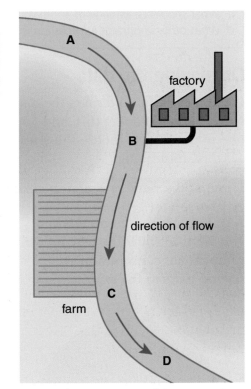

▲ Map showing section of the river and four sampling points.

Questions

c For each letter on the map, A, B, C and D, say why you think that the scientists chose this particular point.
d What variables do you think that the scientists should have controlled or monitored to make sure that it was a fair test?

Presenting the data

The scientists presented their **data** in a table.

The table only shows the average of the measurements. It is often important to see all of the original data in order to see whether there was a wide variation and whether there were any **anomalous** results.

Part of river	Average concentration of nitrates	Average concentration of cyanide
A	2.0	0.02
B	2.5	10.45
C	3.5	8.32
D	2.2	4.78

Questions

a In this investigation, what was (i) the independent variable, and (ii) the dependent variable?
b What is meant by the term 'anomalous results'?
c What should be done with anomalous results before calculating the average?
d What other important information is missing from the table of results?

Displaying the data

Often it is useful to display the data in a graphical form. There are several ways of doing this, and you need to know which is the best one to use. If you choose the wrong one, it can be very unhelpful and at worst misleading.

Three of the most common methods are shown below.

A **pie chart** is often used when you want to show the percentage or fractional contribution of several parts to the whole.

A **bar chart** is often used when you are dealing with variables that are **discrete**, **ordered** or **categoric**.

A **line graph** is often used when you are dealing with variables that are **continuous**.

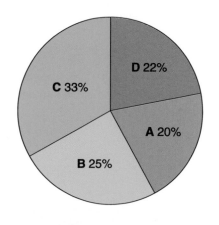

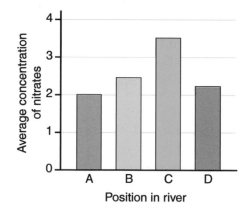

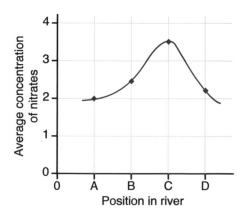

Question

e Which of the three methods of displaying data in graphical form do you think the scientists should have used in this case? Explain your answer.

A **discrete variable** is a variable whose value is restricted to whole numbers.

An **ordered variable** is a variable that can be ranked, e.g. first, second, third, etc.

A **categoric variable** is a variable that can be described by labels, for example by colour or size.

A **continuous variable** is one that can have any numerical value, for example time, speed or length.

Analysing results and drawing conclusions

The scientists issued the following statement:

> Although both pollutants are present in above average quantities, we cannot yet determine which, if either of them, is responsible for killing the fish. This would require more data.

If it were clear that cyanide or nitrates caused the fish to die there would be a **causal** link between one variable causing a change in the second.

Science and society

The managers of the detergent factory looked at the data and made the following statement: 'This has proved that it is not our factory that has caused the problem: it is the nitrates in the water. There is a significant increase in nitrate levels just after the river has passed the farmland.'

Farmer Fred Guest's response to this was: 'Rubbish! The increase in cyanide is far greater than the increase in nitrates. It is the factory's pollution that is killing the fish.'

Sometimes the conclusions that people come to may be influenced by other things and not just the scientific evidence. For example, for many years tobacco companies were accused of misleading the public about the effects of smoking. They might have wanted to do this so that they did not lose sales of their products.

Question

f Which of the following additional data do you think would be most useful to the scientists? Give a reason for your answer.
- A chemical analysis of the dead fish
- Sampling at different points in the river
- More samples at the same points in the river
- More samples over a longer period of time

Questions

g Do you think that either of these claims is reasonable? Explain you answer.
h Suggest one way in which it might be possible to prove that one of the claims is correct.

Question

i Can you think of any reasons why the company operating the detergents factory might have been biased in coming to their conclusion?

Your work as a scientist

These are the sorts of problems and questions that you will have to deal with in your science course. In your practical work you will need to think carefully about all of these points:

- observing
- designing an investigation
- making measurements
- presenting data
- looking for patterns and relationships
- coming to conclusions
- considering the relationship between science and society.

Finally, remember that sometimes it is difficult to collect enough evidence to be able to answer a question properly. There are also questions that cannot be answered by science alone, but need moral or social judgements to be made.

Everything we use comes from the Earth in one way or another. Rock is a material that we can often use straight from the ground. Blocks of stone are cut out of quarries. Limestone is a very common rock, and many old buildings were built using blocks of limestone. If you look closely you can sometimes see fossils in the limestone blocks.

▲ Many old buildings in Oxford are built from limestone.

Chemistry to the rescue

For modern buildings we don't use the natural, raw materials from the Earth directly. You can't build skyscrapers from blocks of limestone. Creative ideas in science and technology have helped us to develop new building materials. The raw materials have to be changed by chemical reactions before we use them.

New buildings such as the Lloyd's building in London are mostly made from concrete and glass. These materials have been used for a long time – the Romans used concrete for buildings nearly 2000 years ago. Over the last 100 years scientists have found out how to make stronger and more versatile concrete. It has become the most important building material of all.

The Romans also used glass – but only in small pieces. Some very old British houses still have tiny glass window panes, which were the best that could be made at the time. But glass technology has improved dramatically over the last 100 years. Now we can make huge sheets to cover our fantastic modern skyscrapers.

What are the raw materials?

Concrete is made from limestone and clay, mixed with sand and gravel.

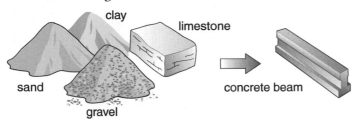

Glass is made from sand and limestone.

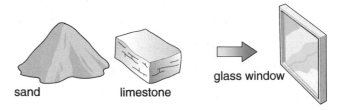

The effects of extraction

Digging out rocks leaves big holes – quarries. Unfortunately, the best limestone is often found in scenic areas, such as the Peak District. Quarries can be ugly places that spoil the natural beauty of the landscape. Of course, once the rock has been removed, the soil can be replaced and trees planted. Some old quarries have been successfully turned into nature reserves or country parks. Elsewhere, old quarries have been used to build shopping malls, such as Bluewater in Kent. More imaginatively, an old clay pit in Cornwall has been turned into the amazing Eden Project.

Quarries are noisy and dirty. Rock is blasted from the quarry face by explosives, spreading dust far and wide. Rock-crushing and sorting machinery rattles away all day and huge lorries rumble down local roads. Most people wouldn't like a quarry next to their home. But we need the limestone for making concrete and other important materials. Quarries also provide employment for local people, which helps the local economy.

▲The Eden Project near St Austell in Cornwall.

Think about what you will find out in this section

How can we make new building materials from the Earth?

How can we decide which new materials are best for the job?

How do we evaluate developments in building materials?

How do scientists use ideas about atoms to explain what is happening when they make new materials?

Tiny particles

If you crushed limestone to a powder, you would have a lot of tiny fragments of limestone. But limestone is made of really tiny atoms. There are about 3 million million million million (3×10^{24}) atoms in 100 g of limestone.

Naming the atoms

Most materials are made from atoms joined by chemical reactions. Limestone is made of three different types of atoms: calcium, carbon and oxygen. They make a chemical compound called calcium carbonate. Glass is made from calcium, silicon and oxygen atoms. These atoms make a chemical compound called calcium silicate.

There are about 100 different kinds of atoms – or elements. Each element has a symbol. Calcium is Ca, carbon is C, oxygen is O and silicon is Si. Calcium carbonate is $CaCO_3$ and calcium silicate is $CaSiO_3$.

> **Question**
>
> a How many atoms would there be in 1 kg of limestone?

> **Question**
>
> b Sulfuric acid has the formula H_2SO_4. What atoms does it contain? And how many of each?

Inside the atom

In their earliest ideas about atoms, people thought of atoms as simple balls. We often draw them like this when we are making simple models of chemical reactions. But 100 years ago, scientists discovered that the atoms themselves were made of even smaller particles.

Atoms have a small central **nucleus**. Whizzing in orbit around the nucleus are the tiny **electrons**. It is the moving electrons that give the atoms their shape. They also control chemical reactions between elements and compounds.

▲ This simple model of an atom shows electrons orbiting the nucleus in a similar way to planets orbiting the Sun in the solar system.

Electrons rule chemistry

When a chemical reaction takes place between atoms, electrons may move from one atom to another or be shared between atoms.

When metals and non-metals combine, the metals give up some electrons and the non-metals take them. Sodium and chlorine form a compound like this called sodium chloride. This is common salt. The sodium and chlorine atoms are stacked up in a **lattice**.

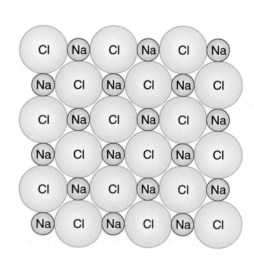

> **Question**
>
> c Describe what happens to some of the electrons when iron atoms react with oxygen atoms to form iron oxide.

When non-metals combine, the atoms share electrons and form molecules. Carbon and oxygen combine like this to form carbon dioxide (CO_2). Hydrogen and oxygen form water (H_2O).

Atoms of elements such as oxygen can also share electrons and make molecules all on their own. The oxygen in the air forms O_2 molecules.

This sharing or moving of electrons from one atom to another makes the **chemical bonds** that hold atoms together in chemical compounds.

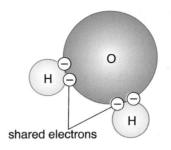
shared electrons

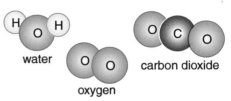
water   oxygen   carbon dioxide

Conserving atoms

When chemicals react, the atoms just rearrange themselves. You always end up with the same number of each type of atom as you started with. You can show this as a balanced equation. For example, calcium oxide reacts with carbon dioxide to form calcium carbonate:

$$CaO + CO_2 \rightarrow CaCO_3$$

Different properties

Different elements have different properties, but there is a pattern to this variation. You can see this in the periodic table. Each vertical column, or group, shows elements with similar properties.

When elements react to form compounds, the compounds have completely different properties from the elements that make them. For example, calcium is a metal, carbon a black solid non-metal and oxygen is a gas. Yet calcium carbonate, made from all three, is limestone.

Questions

d Which of these elements are metals and which are non-metals?
barium (Ba)   selenium (Se)
molybdenum (Mo)   radon (Rn)
lithium (Li)
e Which group contains the most reactive metals?
f Which group contains totally unreactive non-metals?

Group number

1	2											3	4	5	6	7	0
																	He
Li	Be											B	C	N	O	F	Ne
Na	Mg											Al	Si	P	S	Cl	Ar
K	Ca	Sc	Ti	V	Cr	Mn	Fe	Co	Ni	Cu	Zn	Ga	Ge	As	Se	Br	Kr
Rb	Sr	Y	Zr	Nb	Mo	Tc	Ru	Rh	Pd	Ag	Cd	In	Sn	Sb	Te	I	Xe
Cs	Ba	La	Hf	Ta	W	Re	Os	Ir	Pt	Au	Hg	Tl	Pb	Bi	Po	At	Rn
Fr	Ra	Ac															

H

non-metals   metals

Key points

- Atoms have a small nucleus, around which are electrons.
- Atoms and symbols are used to represent and explain what is happening in chemical reactions.
- Elements form compounds by giving and taking electrons or by sharing electrons to form chemical bonds.

Using limestone

Limestone is a common rock with many uses. It is found in many of the most beautiful parts of the country. It is blasted out from the cliffs in large quarries. Some people say that these quarries spoil the natural beauty of the countryside.

Limestone has been used as a building stone for thousands of years. The castle in the photograph was made from blocks taken straight from a quarry. Today, most limestone is used to make something new. It is changed by chemical processes into other useful products such as cement, or is broken up for roads or concrete.

Limestone is a very important raw material. About 90 million tonnes of limestone are quarried every year in Britain alone. We use the limestone:

- for buildings or road 'chippings' (66 million tonnes)
- to neutralise excess acid in lakes or soils (1 million tonnes)
- to make cement (15 million tonnes)
- in other industrial processes (8 million tonnes).

Question

a Use the data to draw a pie chart showing the uses of quarried limestone.

Making new materials from limestone

Limestone is the compound calcium carbonate. Its chemical formula is $CaCO_3$. This means that for every calcium atom there is one carbon atom and three oxygen atoms. All carbonates have one carbon atom and three oxygen atoms arranged like this in a lattice.

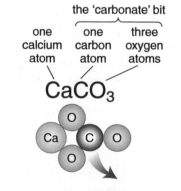

the 'carbonate' bit

| one calcium atom | one carbon atom | three oxygen atoms |

$CaCO_3$

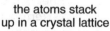

the atoms stack up in a crystal lattice

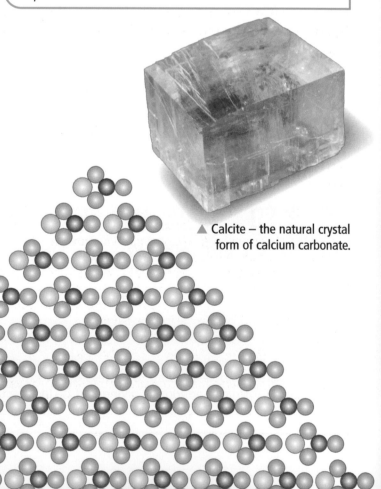

▲ Calcite – the natural crystal form of calcium carbonate.

If limestone is heated strongly, the compound breaks up and the atoms are rearranged.

- The carbon atom takes two oxygen atoms to form carbon dioxide.
- The calcium atom is left with just one oxygen atom. This is calcium oxide, which is also called **quicklime**.

calcium carbonate $\xrightarrow{\text{heat}}$ calcium oxide + carbon dioxide

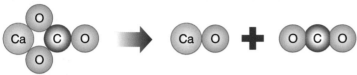

Question

b Copy and complete this equation. Make sure it is balanced.
$CaCO_3 \rightarrow CaO +$ _____

Breaking down a compound by heating it like this is called **thermal decomposition**. As in all chemical reactions you end up with the same atoms – they have just been rearranged. So the products have the same mass as the reactants. You can see the conservation of mass in these examples:

100 g of calcium carbonate gives 56 g of calcium oxide and 44 g of carbon dioxide

or

20 g of calcium carbonate gives 11.2 g of calcium oxide and 8.8 g of carbon dioxide

Reacting quicklime

Quicklime is a very strong alkali. It reacts with water to form calcium hydroxide, also known as **slaked lime**. A lot of heat energy is given out in this reaction.

calcium oxide + water $\rightarrow$ calcium hydroxide + heat energy

$\quad CaO \quad + \quad H_2O \rightarrow \quad Ca(OH)_2$

Farmers have used slaked lime for centuries as a simple fertiliser to make their soil less acid. It also helps to break up the soil so that plants can grow well. Sometimes lakes are polluted by acid rain. Slaked lime can be used to get rid of the acid there as well.

Questions

c If 28 g of calcium oxide reacts completely with 22 g of carbon dioxide, what mass of calcium carbonate would you get?

d In the thermal decomposition of calcium carbonate, the solid calcium oxide left behind weighs less than the calcium carbonate. Why?

e Iron carbonate ($FeCO_3$) breaks down just like calcium carbonate. What new chemicals would you get if you heated this?

f Copper carbonate ($CuCO_3$) breaks down in the same way to give copper oxide (CuO). Write a balanced equation for this reaction.

g Quicklime has to be handled very carefully. Explain why.

h Mixing 56 g of quicklime with water makes 74 g of dry slaked lime. What mass of water has reacted with the quicklime to make this?

Key points

- The formula of a compound shows the number and type of atoms in it.
- In chemical reactions, atoms are rearranged. No atoms are lost or made, so we can represent reactions using balanced equations.
- Limestone can be used as a building material or as a raw material for new products.
- Limestone ($CaCO_3$) and other carbonates can be broken down by thermal decomposition.

New rocks from old

Rock or 'artificial rock'?

These statues look very similar. One took a sculptor many weeks to make by shaping the limestone. The other was made by simply pouring concrete into a mould. Concrete is like an 'artificial rock'.

Making mortar, cement and concrete

Houses in many parts of Britain were traditionally built from bricks stuck together with **mortar**. The mortar is made by mixing slaked lime (made from quicklime) with water to make a thick paste. This dries and sets hard between the bricks, holding them together. Over time it reacts with carbon dioxide from the air and turns back to limestone. In many old houses this mortar has turned soft and crumbly. The bricks have to be 're-pointed' with fresh cement to stop the old mortar washing away in the rain.

Today, most quarried limestone is used to make **cement**. The limestone is heated with clay in a big oven and thermal decomposition occurs. The oven is called a rotary kiln because it keeps turning to mix everything up. The roasted product is then ground to form a light grey powder. This is cement. It forms a paste with water which soon hardens back to rock, like mortar, but the chemical reaction is more complicated. A series of chemical reactions take place involving calcium, silicon, oxygen, iron and aluminium. The new compounds form tiny crystals which interlock tightly to make a hard, rock-like material.

Cement is used to stick bricks together today instead of simple mortar. But most cement is mixed with sand, gravel and water to make **concrete**. This is cheaper and stronger than pure cement.

> **Question**
>
> **b** Every tonne of limestone that is turned into cement releases 440 kg of carbon dioxide into the atmosphere. Suggest one disadvantage of modern cement over old-fashioned mortar.

Concrete forms a thick liquid when first mixed, and can be poured into any shape. Slow chemical reactions make it set after a few hours and eventually it becomes rock hard. It is used to make roads, bridges and the frameworks and foundations of buildings.

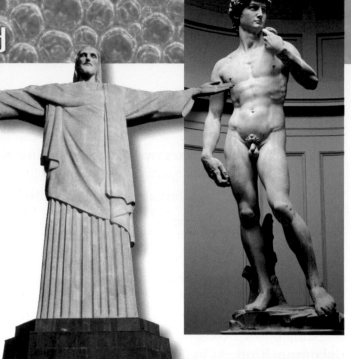

> **Question**
>
> **a** Heating limestone to make quicklime releases lots of the greenhouse gas carbon dioxide into the air. Why is this not a problem if the quicklime is turned into slaked lime and used to make mortar?

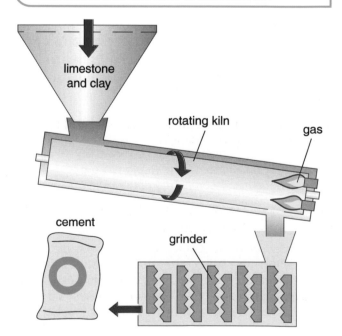

Making glass

Small pieces of glass have been used for thousands of years. But these days glass is a very important building material. Many modern buildings are completely covered in glass. Glass is also made from limestone. This time the limestone reacts with another common raw material – sand.

Pure sand is usually grains of silicon dioxide. Glass is made by heating very pure sand with limestone and a little soda (sodium carbonate). The chemicals react, melt, and carbon dioxide is released. When this liquid cools, it forms a hard, but brittle, transparent material – glass.

calcium carbonate + silicon dioxide → calcium silicate + carbon dioxide

Question

c Copy and complete this to make a balanced equation.
$CaCO_3 + SiO_2 \rightarrow$ _____ + _____

Smart glass

Sunlight can damage your eyes, so it is a good idea to wear sunglasses on a very sunny day. But what can you do if you already wear glasses? Smart glass to the rescue! Photochromic glasses contain special dyes in the glass of the lenses which darken in sunlight – but go clear again when the sun goes in.

Question

d The first photochromic glasses took a few minutes to clear when the light levels dropped. Why was this a problem for motorists driving into tunnels?

As this glass gets cheaper, it could be used for windows in sunny offices or classrooms instead of blinds.

One new form of glass even has a special coating that automatically cleans itself. This uses the energy from the ultra violet rays in sunlight to break down the dirt. Rain then simply washes it away!

Question

e Suggest some of the problems of working in a glass office building and of maintaining such a building.

▲ The Swiss Tower, London (better known as the London Gherkin).

Key points

- Limestone can be used as the raw material for quicklime, slaked lime, cement and concrete.
- Limestone is also the raw material for glass, along with sand.

Concrete: beauty or the beast?

When architects designed the Royal Festival Hall in London over 50 years ago, they wanted to show off their new building material. They left the concrete exposed for all to see. Some people like it, but many find the buildings ugly. It doesn't help that the gleaming white concrete soon gets streaked and dirty in polluted London.

The Bahai Lotus temple in Delhi is also made of concrete. Concrete is so versatile you can build whatever you like from it. The only limit is your imagination. The world is dotted with brilliantly imaginative buildings that rely totally on concrete.

▲ The Royal Festival Hall was built in 1951 as part of the Festival of Britain.

Getting the best out of concrete

Building with quarried stone was a slow and labour-intensive process. Each block had to be carefully cut to shape and fitted into place by an expert craftsman. This made it slow and very expensive.

▲ The Bahai Lotus temple in Delhi, India, which was built in 1980.

In comparison, building with concrete is quick, cheap and easy.

- Cement is relatively cheap to produce on an industrial scale.
- Powdered cement is easy to store and transport.
- Sand and gravel can be found almost anywhere and are cheap materials. You can even recycle building rubble or ash from power stations to use as gravel.
- Liquid concrete can be easily mixed on site.
- Liquid concrete can be moulded into any shape you want: floor, girders, domes, ornaments.
- Once it sets it really is rock hard.

Concrete has one problem that it shares with quarried limestone. It is very strong if you squash it, so it can support very large buildings such as skyscrapers. But it is brittle, so if you stretch or bend it, it can crack easily. To overcome this problem, concrete beams or girders are reinforced with steel rods. These stop the concrete from stretching and cracking.

Questions

a Explain why it would be hard to build a limestone block skyscraper.

b Old buildings often had wooden beams over doors and windows to support the weight of stone or brick above. The wood was strong enough and could bend slightly without cracking. Today these beams are more likely to be made of concrete.
   (i)   What advantage does wood have over simple concrete for this use?
   (ii) How could the concrete be made stronger, to overcome this problem?
   (iii) What's the disadvantage of using wood if you want your building to last a very long time? (Hint: What happens to wood eventually?)

Getting the best out of glass

Glass is a wonderful material. Glass-fronted offices give a feeling of space because you can see outside. Many people like to feel the warmth of the sun through the glass, and this can help to cut down heating costs as well. But this can be a problem in the summer, when the building could trap heat like a greenhouse and become too hot to work in.

Ordinary glass has one disadvantage. It is brittle and breaks into razor-sharp pieces. Fortunately, advances in technology mean glass can be made much tougher. One way is to heat it and then cool it quickly. This makes the glass five times as strong. Toughened glass is used in places where ordinary glass would shatter, such as car windscreens.

Car windscreen glass still shatters in a crash, but it breaks into small but chunky pieces that are not so sharp as ordinary glass slivers. More modern versions have an internal plastic layer that holds all the pieces together. As technology improves, glass keeps getting better and better. But even the best glass will shatter if there is an explosion or a building is shaken by a powerful earthquake.

Questions

c Suggest two ways to stop glass buildings overheating in summer.

d Suggest what problems there might be in winter.

e Suggest some places other than car windscreens where toughened glass ought to be used.

f Suggest a reason why we don't use toughened glass all the time.

g Describe how modern car windscreens have made it easier to replace a windscreen and clean up after a crash.

Key points

- Concrete is a strong, inexpensive and easy to use material for building. It is very versatile but needs to be reinforced with steel to stop it cracking if used for beams or girders. Some people think concrete buildings are ugly.
- Glass-fronted buildings can be good to work in as they are light and airy, but they can overheat in summer if not designed well. Broken glass is also very sharp and dangerous.

Useful metals

Technology relies on the strength of metals for constructing everything from aeroplanes to tin cans. Thousands of kilometres of electric cables snake across the country to bring us the power we need for our electrical equipment. We are very dependent on metals.

The useful properties of metals

It is the useful properties of metals that make them important.

- Metals are strong and hard, which makes them good structural materials. We use them to build machines, bridges and the frameworks for large buildings.
- Metals are easy to shape. Car body panels are pressed out of sheet steel.
- Metals have high melting points. This means that you can build engines that won't melt when you use them.
- Metals also conduct heat and electricity. They are used in electrical wiring and as 'heat sinks', to conduct heat energy away from microprocessors, to stop them overheating.

Are there problems with metals?

Some properties of metals are less useful.

- Some metals such as gold are very rare and so very expensive.
- Some metals are expensive because they are difficult to extract.
- Pure metals are too soft to be of much use, so scientists have worked out how to make **alloys** with improved properties.
- Metals corrode. We spend money and effort getting iron from the rocks – but leave any iron object lying about in the rain and you end up with a pile of rust.
- Metals pollute. Waste metal can corrode and get into the soil and groundwater. This can kill plants and animals. If it gets into the food chain it can harm us too.

▲ The *Pacific Princess* is a typical metal-hulled ship.

And what if we run out?

Modern technology relies on metals. But at present rates, using today's mining techniques, we may run out of many important metals over the next 100 years. What can we do to stop this happening? Some scientists are trying to find new sources of metals and new ways of extracting them. Others are looking for better ways to recycle the metals we have.

Steel or wood?

When builders and architects are looking at what materials to use, they have to evaluate the materials by looking at their properties, advantages, disadvantages, availability and costs.

In the USA, many houses are built using wooden beams, which cost 10% less than the steel equivalent. Overall, similar houses cost 10% more if built using steel. Wood is cheaper because it is a natural, renewable raw material. Steel has to be manufactured from finite iron ore sources. Wood can be cut and shaped on site, but steel beams arrive ready made in fixed sizes.

One of the disadvantages of wood is that it may rot in time. Steel is longer lasting, though it can rust if it is not protected. Although wood is strong enough for buildings of small to moderate size, it is not strong enough for large buildings. Steel beams are strong enough for even the tallest skyscraper.

▲ The Tamar road bridge.

▲ Japanese bullet train.

Think about what you will find out in this section

How do we get metals from their ores?

How are the most commonly used metals iron and steel made?

How can science help to make recycling easier?

How do we improve metals to make them better suited to their uses?

How can we evaluate our use of metals as structural and smart materials?

Where do metals come from?

Metals are found in chemical compounds in the rocks of the Earth. Some metals are quite common. A field of mud contains tonnes of aluminium, and a lot of iron too. The trouble is the metals are far too difficult to extract. Fortunately, natural processes sometimes concentrate metals in certain rocks, which makes it easier and cheaper to extract them. Rocks like this are called **ores**.

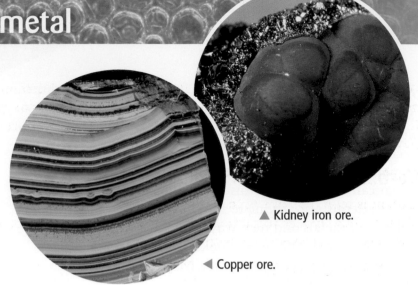

▲ Kidney iron ore.

◄ Copper ore.

Which metals are common?

Aluminium and iron are two metals we use a lot. As you can see from the table, they are also very common in rocks. But some other important metals, such as copper and gold, are really very rare.

Metal	Percentage of the rocks of the Earth's crust
aluminium	7
iron	4
magnesium	2
copper	0.0045
tin	0.0002
gold	0.0000005

> **Question**
>
> **a** How much more common is iron than gold? Choose from 8 thousand times, 80 thousand times, 8 million times or 80 million times.

Iron is king

Iron is the metal we use most. In 2004, the global production of iron topped 1 billion tonnes for the first time. That is 20 times as much iron as all the other metals put together. Production on this scale helps to make iron cheap compared to aluminium, even though aluminium is more common. But iron is also cheaper because of the way it is extracted from its ore.

Metal	Price per tonne (spring 2005)
aluminium	£1800
copper	£3300
iron	£300
tin	£8000

Reactivity and metals

Gold is a very unreactive metal. It does not react with other elements. It may be rare but, if you are very lucky, you could find a nugget of pure gold!

> **Question**
>
> **b** Explain why rare gold was discovered before common aluminium.

▶ Gold was well known to the Ancient Egyptians.

Aluminium is a very reactive metal. Because of this, the aluminium atoms are tightly combined in compounds with other atoms such as oxygen. Mud contains plenty of aluminium, but you cannot easily extract it. You never find aluminium as a pure element in the Earth's crust.

Carbon reduction and oxidation

Carbon is more reactive than metals such as iron and copper, so it can displace these metals from their compounds. We can use carbon like this to get less reactive metals from their ores.

Question

c Explain why you can't get aluminium from its ore by this method.

Metal ores are usually oxides. When we react the ore with carbon, the carbon combines with oxygen from the ore to form carbon dioxide. The carbon is oxidised. This is an **oxidation** reaction.

The metal oxide ore has its oxygen taken away. The oxide is reduced to the metal. This process is called **reduction**. Reduction is the chemical opposite of oxidation, and the two reactions always go together. For example, with copper oxide:

reduction

copper oxide + carbon → copper + carbon dioxide

oxidation

$$2CuO + C \rightarrow 2Cu + CO_2$$

Metals have been made using this method for thousands of years. Originally the carbon was in the form of charcoal. Today coke (made from coal) is used instead.

The Wealden Forest in Kent doesn't have any trees now. They were all chopped down to make charcoal for iron production at the start of the Industrial Revolution.

▲ There is plenty of aluminium in this mud! But how could you tell?

most reactive

↑

magnesium (Mg)

aluminium (Al)

carbon (C)

iron (Fe)

tin (Sn)

copper (Cu)

gold (Au)

least reactive

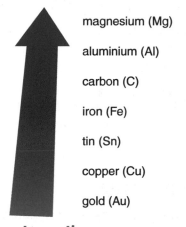

Questions

d Tin oxide (SnO_2) reacts with carbon to give tin and carbon dioxide. Write this as a word equation, showing the oxidation and reduction arrows.
e Copy and complete the equation. $SnO_2 + C \rightarrow$ _____ + _____
f Write a simple word equation for the reduction of iron oxide by carbon.

Key points

- Ores are found in the Earth and contain metal compounds from which the metal can be extracted. Some metals are more economic to extract than others.
- Metals such as copper can be extracted by carbon reduction.

Wanted by all

Iron is the most widely used metal, so we need to produce plenty of it to supply the world. Over the last 300 years scientists have refined the carbon reduction method to make it very efficient. But iron is not as easy to extract as copper, tin or other less reactive metals.

Iron plays hard to get

Carbon reduction reactions need a kick-start of energy to get them going. For copper and tin, the heat from a simple fire will do it. But for iron, much more energy is needed to get the reaction started. You can make charcoal burn at a very high temperature by providing more oxygen. Fanning a barbecue provides more oxygen and makes it hotter. This idea is used in a blast furnace to reach the very high temperatures needed to reduce the iron ore.

▲ Molten iron pours from a blast furnace.

Question

a Blacksmiths used bellows for pumping air. Why was this better than fanning the charcoal?

Inside the blast furnace

Iron ore (haematite, a type of iron oxide), coke (carbon) and limestone are tipped in at the top of the blast furnace. The main reaction is the reduction of iron oxide and the oxidation of carbon.

iron oxide + carbon → iron + carbon dioxide

Question

b Complete the balanced equation for this reaction:
$2Fe_2O_3 + 3C \rightarrow$ ___Fe + ___CO_2

Most iron ore used today is only about 50% iron oxide. This will produce about 350 kg of iron per tonne (1000 kg) of ore. But the impurities from 1 tonne of ore will react with the limestone to make just under 1 tonne of waste material called slag.

Question

c *Here is a working 'recipe' for a blast furnace using the 50% ore. How much 'missing mass' is there? How can you account for it?*

Cast iron – carbon makes the difference

The iron that comes from the blast furnace is called cast iron. It's 96% iron with about 4% carbon (and some other impurities). The impurities make it very hard but quite brittle. If you hit it with a heavy hammer it would shatter. In the past, cast iron was used to make everything from manhole covers to bathtubs. It was used to make girders for the bridge at Ironbridge in Shropshire. Nowadays it is thought to be too brittle for most uses.

Question

d *Ironbridge in Shropshire was fine for horses and carts. Why couldn't this cast iron bridge be used for modern heavy lorries?*

Pure iron can be made by bubbling oxygen through the molten iron to burn out the carbon. But pure iron is very soft – too soft to use for construction. Fortunately it can be made much harder by leaving just a little carbon – just 1% or so. This new material is hard and tough! This alloy of iron is called steel. All the iron we use today is in this form.

The price of steel

The world price of metal changes over time, depending on what it costs to extract metal from its ore, the percentage of metal in its ore and its availability, as well as the supply and the demand for it in the world market.

Questions

e *Over the last few years, China and India have started to industrialise very rapidly. What effect do you think this has had on the demand for steel?*

f *What effect do you think that had on the price of steel on the world market?*

For every tonne of ore add:
850 kg limestone
and
50 kg coal

You should get:
350 kg iron
and
910 kg slag

Key points

- Iron is extracted from its ore in a blast furnace.
- Impurities make cast iron very brittle but pure iron is very soft.
- Iron is most useful when converted into steel with the addition of 1% carbon, making an alloy.

Why are alloys harder than pure metal?

In a pure metal all the atoms are the same, stacked up in a regular way. They are strongly held together, but the layers of atoms can slide over one another if a force is applied. In an alloy, a few different atoms have been added. These are not the same size as the rest.

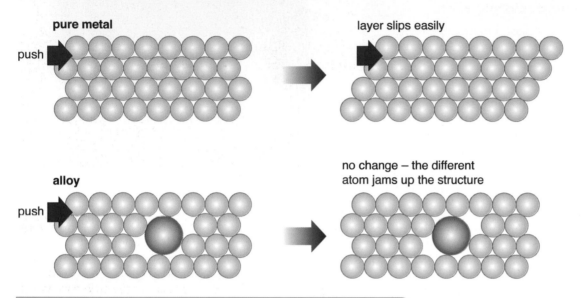

Question

a Use the diagram to explain how the different atoms stop the layers from sliding easily and make the alloy harder.

The best steel for the job

Adding carbon to iron makes it harder because the layers of atoms can't slide so easily. But adding too much carbon makes the lattice weaker. It becomes brittle. The graph shows how steel hardness and strength change with just a small change in the amount of carbon.

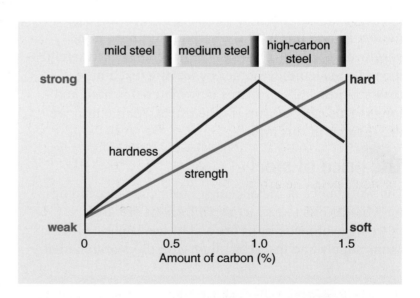

- Mild steel (<0.5% carbon) is quite soft. In thin sheets it can be pressed and moulded.
- Medium steel (0.5–1.0% carbon) is harder. It is strong enough to use for hammers.
- High-carbon steel (1–1.5% carbon) is hard enough to give a good cutting edge. It is not as strong as medium steel and can be brittle.

Question

b Which type of steel would be best for: (i) scissors and chisels, (ii) making car body panels and (iii) making hammers or spanners? Explain your answers, using the graph to help.

Special steels that don't corrode

Sometimes other metals are added to steel to give new alloys with special properties. Nickel is added to make a very tough steel. Tungsten is added to make a very hard steel. Adding 15% or so of chromium has a different effect. This stainless steel does not rust.

Question

c Stainless steel is perfect for cutlery, razor blades and the steam pipes of power stations. Explain the problems that corrosion would cause in each of these cases if stainless steel wasn't used.

Smart alloys

Pure copper, aluminium and even gold would be too soft to use on their own, so they are also mixed with other metals to make useful alloys which are harder.

- Gold is mixed with a little copper for jewellery.
- Aluminium is mixed with magnesium and a little copper for building aeroplanes.

Question

d Brass is made from copper and zinc. Why are bolts and hinges made from brass and not copper? Explain the difference in properties.

Metal scientists can now make fantastic new alloys with improved properties. At their simplest, some are *superelastic.* They can be bent and twisted without damage – great for making glasses that don't break if you sit on them! But some alloys have even smarter properties. They can be bent at low temperatures but they snap back to shape if they warm up. These shape memory alloys are made from metal combinations such as nickel/titanium or copper/nickel/aluminium.

Many people who suffer from heart disease have blocked or collapsed arteries. A wire grid can open the arteries up, but how can it be put in place? A grid made from a shape memory alloy can be cooled and squashed to fit in easily. As it warms up in the body, it snaps back to size and opens the artery.

collapsed artery   squashed grid in place   warms and expands

Question

e Orthodontists have to screw brace wires tight every few weeks to pull teeth into line. This is difficult and wires can sometimes be made painfully tight. Stretched cold 'smart wires' could be used which would contract in a much more controlled way when they warmed. How would this be better?

Key points

- There are many different types of steel, with different properties depending on the carbon content.
- There are many other alloys made with other metals to give useful properties.
- Smart alloys can return to their original shape after being deformed.

A hole in the ground

Copper is a very important metal, but there is only a thousandth as much of it in the Earth as there is iron. So when we find a big body of concentrated copper ore we just keep on digging till we've got it all out. The Bingham Canyon mine in Utah, USA, is now nearly 1 km deep!

What's so good about copper?

Copper is one of the **transition metals** from the central block of the periodic table. The transition block contains the 'everyday' metals. They are useful materials to make things from as they are hard and strong – unlike the metals in Groups 1 and 2. They can be bent or hammered into shape and will keep their new shape. Like all metals, they also conduct heat and electricity.

Copper is particularly useful as it is such a good electrical conductor. It is also soft and bendy, so it's great for electrical wires. Copper is not very reactive and so does not corrode like steel. This makes it useful for copper pipes for plumbing.

Getting out the metal

Digging copper ore out of the ground is just the start of the extraction process. The copper is usually in the form of copper sulfide crystals scattered through the rock. In the past, this was converted to copper oxide and then the copper was extracted from the ore by reduction.

> **Question**
>
> **a** In ski resorts, copper sheeting is sometimes used for roofs. Suggest two reasons why copper might be used here instead of cheaper steel.

▶ The Bingham Canyon copper mine, Utah, USA.

There are just a few major copper mines dotted around the world – in the USA, South America, Africa and some islands around the Pacific. The ore only contains about 1% of copper. Separating out the copper by reduction is too expensive, so nowadays other methods are used. One method is called leaching. Acid is sprayed onto the rock. Soluble copper compounds dissolve out of the rock and the solution is collected. The copper is then removed by **electrolysis** – breaking the compound apart using electricity.

We use large amounts of copper and the copper mines will soon be exhausted. Many old mines are now reworking their old waste tips to get out more of the copper by leaching.

◀ 'Peacock' copper ore.

So what's the future?

Leaching can get just a fraction of 1% of copper from the rocks, but scientists are working hard to find ways of extracting smaller and smaller fractions. One new method involves using special bacteria that 'eat' the copper from the rock.

Another problem is that even if we find a new source of copper, environmentalists have started to object to mining companies ripping great holes in the Earth! Open-cast copper mines are very large and very ugly.

A new leaching method that does not affect the environment so much involves drilling down to the ore. Acid is then pumped down to the ore and comes back up to the surface rich in copper for processing. You don't need to dig big, ugly holes in the ground and as long as you can collect all of the solution that comes back up, the environment is not polluted.

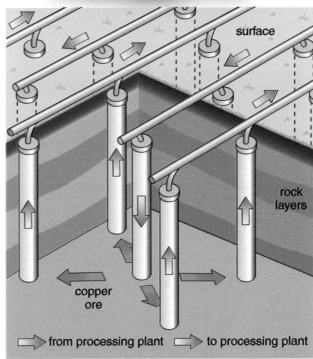

But will we still need copper?

Copper is mainly used for wiring. But new communications systems use laser light, shone along glass fibres, instead of sending information along electrical wires. The most modern systems are now completely 'wireless'. So perhaps we won't need so much copper in the future after all!

Questions

c What will happen to the price of copper if demand falls?
d What will happen to the mines?
e What would be the wider effects on the society, the economy and the environment of mining areas if copper was no longer needed?

Question

b A mining company wants to develop a new open-cast copper mine near you. What objections would you have to the plan and what alternatives could you suggest?

Key points

● Transition metals are good conductors of electricity and heat and can be bent into shape – properties which make them very useful.
● Copper is extracted by electrolysis but other ways are being developed to use ores with low copper content and limit environmental impact.

The lightweight champions

Useful metals with low densities

Steel may be fantastic for cars, trains and ships but its high density makes it useless for aircraft. An aircraft made from steel would weigh twice as much as a modern plane. It would never get off the ground!

Aluminium is not as strong as steel but its density is very much lower. This combination is just right to make a plane that is both strong enough *and* light enough to fly. Without aluminium there could be no commercial airlines.

► Commercial aircraft rely on lightweight yet strong aluminium.

▼ High-performance fighters need even stronger (but more expensive) titanium.

Metal	Density (g/cm³)	Strength	Melting point (°C)
pure aluminium	2.7	low	660
steel	7.7	high	1540
titanium	4.5	high	1670
Duralumin alloy	2.8	medium	600

Supersonic fighter jets need to fly so fast that the wings would get hot enough to melt aluminium. A new 'supermetal' was needed that was strong and had a low density but a high melting point. Titanium fitted the bill perfectly. Titanium is as strong as steel but its density is low enough to keep the weight down. An added bonus is that its melting point is higher than that of steel – high enough to withstand the frictional heating caused by supersonic flight.

Question

a Pure aluminium is not used for aircraft. It is combined with magnesium and a little copper to make Duralumin alloy. Why do you think that is?

The cost of extraction

Aluminium is more reactive than carbon. You can't use the carbon reduction method to get aluminium from its ores. Aluminium ore has to be melted and then split apart using large amounts of energy in the form of electricity. Aluminium plants have to have their own power station to provide this. This is a very costly process which makes common aluminium much more expensive than iron.

Question

b Some of the first aluminium plants were built in the mountains of Scotland. What relatively cheap method of making electricity do you think they used? (Hint: Look at the photo.)

Titanium is also more reactive than carbon and so can't be made by carbon reduction. But titanium is less common than aluminium and its concentrated ores are harder to find. It is also much harder to extract from its ore. This means that titanium is much more expensive even than aluminium. It is only used where high performance is more important than cost.

Question

c Some lower performance supersonic planes just use titanium for the nose-cone and leading wing edges. Suggest two reasons for this.

Resisting corrosion

Aluminium and titanium share another useful property. They both resist corrosion well. That is why aluminium foil stays shiny. Titanium resists corrosion much better than aluminium and stainless steel. It can be used safely where even stainless steel would corrode away, for example in nuclear power stations – or inside the human body!

Question

d What would happen in time if the hip replacement shown on this X-ray photograph was made from steel?

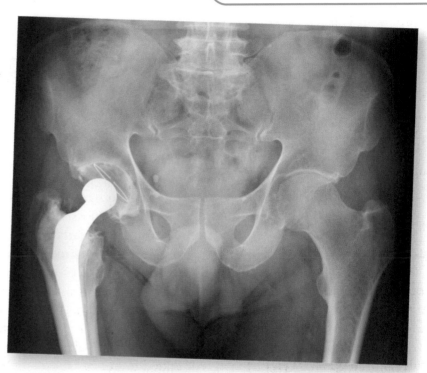

Recycling metals

The table shows which metals we **recycle** in Britain.

Metal	% recycled in Britain
aluminium	43
copper	45
iron	46
lead	61
tin	30
zinc	14

Questions

e Which metal is recycled the most?
f Aluminium is the commonest metal in the Earth's crust. Explain why it is particularly important to recycle aluminium despite this.

Key points

- Aluminium and titanium have two very useful properties: low density and resistance to corrosion. However, their extraction is expensive as it requires a lot of energy.
- We need to recycle metals as much as possible because the amount of ore is limited and extraction is costly in terms of energy and its impact on the environment.

Scarce resources

Earth is like a spaceship. We all live and get everything we need from it. There are nearly 6 billion people on the planet and we are consuming Earth's resources at an alarming rate. Fossil fuels like oil and gas will soon run out. Some metal ores will become scarce within the next 100 years or so.

Metal	Proven reserves will last until ...
tin	2030
copper	2030
tungsten	2050
aluminium	2050
nickel	2100

Questions

a Which metals might run out in your lifetime using today's mining techniques?

b Iron is unlikely to run out in the near future. But steels used for machinery or other high-performance uses are alloyed with metals such as vanadium and manganese that will run out. What problems would society face if we ran out of metals for these important alloys?

All mined out

Some optimists think that using up metal ore resources is not really a problem. As our technology improves it becomes possible to extract metals from poorer quality ores. The waste tips from early mines have often been successfully reworked. Also, if a metal starts to run out, the price goes up, so old mines that had to close when prices were low might reopen when prices rise.

Are we exploiting poorer countries?

Some countries are rich in mineral resources, while others have few. Industrial societies are buying up more than their fair share of the global resources. Sometimes it is because the countries of the developed world have used up their own resources. But even if the ore is still available in developed countries, it is often cheaper to mine in developing countries because the wages of the miners and other costs are so low. This does have some benefits as it provides jobs for people in the developing countries. But sometimes the working conditions are not very good and health and safety rules are not applied as strictly as they would be in the West.

Zambia is a poor country. Its economy relies on exports of copper from its huge copper mines. Forty years ago Zambia was encouraged to take out huge loans from the World Bank to develop these mines. They said it would raise the standard of living for all Zambians.

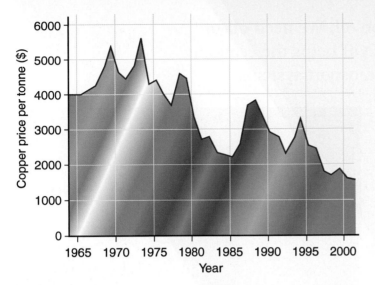

Questions

c What happened to the price paid for copper in the late 1970s?
d What effect do you think this had on the people and economy of Zambia?
e Do you think the advice from the World Bank was good or not? Explain your answer.

The economics of recycling

Recycling costs money. One of the problems is that waste material may well have lots of different metals mixed up. It is only economically viable at the moment for metals that are easy to separate such as iron and steel, or expensive metals such as gold. As technology improves, perhaps we will be able to do this for all metals.

Alucycle is a large company that produces 1 million tonnes of aluminium goods every year. It has its own plant for making aluminium from aluminium ore, but also uses recycled aluminium. The energy used for making the metal from its ore costs £100 per tonne. The energy used recycling aluminium costs just £5 per tonne. The company currently uses 40% recycled aluminium in its products.

Questions

f Calculate Alucycle's energy costs for its annual production of aluminium goods.
g The company hopes to increase the amount of recycled aluminium it uses to 50%. How much money a year would it save on energy costs?

The future is out there …

If we eventually colonise Mars, it will be too expensive to ship out our raw materials from Earth. Mars may have its own metal ores, but just outside the orbit of Mars lies the asteroid belt. This is made up of millions of orbiting chunks of rock. Many are several kilometres across – and about half of them are made entirely from iron and nickel, just like the core of the Earth. It is a space technologist's dream that one day we will be able to mine these 'flying mountains' of metal.

Question

h List a few of the problems you would have to overcome before you could set up a mining operation on an asteroid.

Key points

- Mining, extracting and recycling metals have economic, social and environmental impacts.
- The supply of metals on the Earth is finite so we need to develop ways of using and recycling them effectively.

Our shiny new technological world relies on oil. We need oil to fuel our cars, machines and power stations. Oil helps the economy grow so we can afford new buildings and transport systems.

▲ Oil money pays for developments like this in Bahrain.

▲▼ But oil can do great damage if handled badly…

Environmental impact

But there is a price to pay for economic success. The catastrophic pollution caused by burning oil wells after the first Gulf War is just an extreme example of the environmental hazards posed by oil.

Coastal areas from Alaska to Cornwall have been devastated by oil spills from wrecked tankers over the years. Sea birds are the most obvious casualties, but the whole ecosystem suffers. Oil usually contains a little sulfur. If this is not removed before burning, sulfur dioxide gets into the air and causes acid rain. The carbon dioxide from burning could be causing global warming.

Human impact

Around 50 years ago only a minority of families owned cars. Today, almost everybody expects to own a car – often two or three per family. Exhaust gas pollution causes the brown haze that hangs over many cities on hot summer days.

About 50 years ago asthma was uncommon. Today 15% or so of children suffer from asthma, and need to use chemical inhalers to help them breathe when they have an attack. Attacks happen more often on high pollution days.

The rise in asthma seems to match the rise in air pollution. Does that mean air pollution causes asthma? The graph alone does not prove anything – it could just be chance or some other factors that have changed with the change in lifestyle over the last 50 years. Research suggests that pollen and even thunderstorms can also trigger asthma attacks, but air pollution *may* trigger the problem in the first place. Science has not yet proved conclusively that this pollution causes asthma. Even so, the pollution levels are monitored carefully and 'high pollution' alerts go out with the weather forecasts to help warn asthma sufferers.

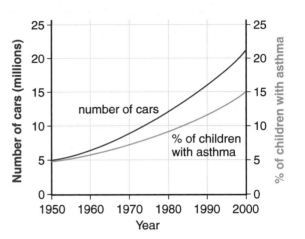

Economic impact

North Sea oil has been great for Britain. The production of over 2 million barrels of oil per day for the last 30 years has helped to keep our economy healthy and created many new jobs. But not all countries have been so lucky. Nigeria is the biggest oil-producing country in Africa. Despite producing roughly the same amount of oil as Britain since the 1970s, the standard of living for ordinary Nigerians has been declining steadily. Most of the oil money has gone to a few rich people while other industries have been neglected.

Political impact

At the rate we are using oil, it will all run out within your lifetime. The economy of the USA powers the global economy. It was built on vast supplies of cheap oil, but it has almost exhausted its own oil and now needs to buy in oil from the rest of the world. Information about oil reserves shows why the USA takes such a political interest in the Middle East. Many people suggest that the recent Gulf Wars in Kuwait and Iraq had more to do with oil than democracy. Further wars could break out in the future in other parts of the world over the last, dwindling oil supplies.

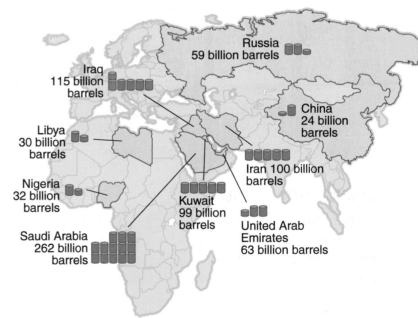

▲ Oil reserves of the major oil-producing countries in and around the Middle East in billions of barrels of oil.

Think about what you will find out in this section

What is crude oil made of?

How can science help to overcome any pollution problems caused by oil?

How can we make useful products from crude oil?

How can science help to develop new energy sources for the future, when oil runs out?

What environmental problems are linked to the use of crude oil?

How did crude oil form?

Crude oil is an ancient biomass. If dead plants or animals get buried quickly, they are broken down by microbes that do not need oxygen. Oil and natural gas are formed by this process. The oil and gas trapped in rocks has formed in this way over millions of years.

Question

a Why are oil and gas called fossil fuels?

What is crude oil made of?

Crude oil is a mixture of many different compounds. Mixtures are not chemically combined, so each compound has its own properties. On their own, they may be runny liquids, thick liquids, solids or gases. Mixed up together they are a thick, gooey black liquid. As a mixture, crude oil is useless. The different compounds must be separated out before they can be used.

Question

b Ink is a mixture. Suggest how you could separate out the pure water in ink.

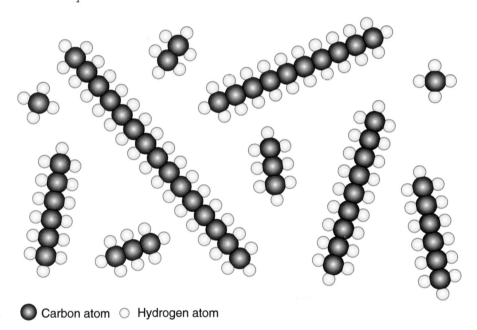

● Carbon atom ○ Hydrogen atom

What kind of compounds?

Most of the compounds in oil are made from two types of atom only: hydrogen and carbon. Compounds like this are called **hydrocarbons**. The hydrocarbons shown here have molecules of different sizes.

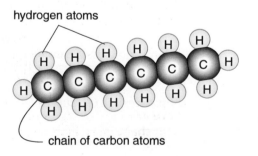

hydrogen atoms

chain of carbon atoms

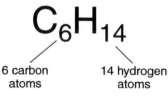

This has 6 carbon atoms. It can be written as:

$$C_6H_{14}$$

6 carbon atoms   14 hydrogen atoms

Carbon atoms can form chains. Big molecules have long chains and small molecules have short chains. The molecules are often drawn like this, showing how the atoms are joined by single chemical bonds. If you look closely, you will see that every carbon atom has four bonds. Two bonds link the carbon atom into the chain and two are linked to hydrogen atoms. At each end, there is an extra hydrogen atom.

$$H-\overset{\overset{\displaystyle H}{|}}{\underset{\underset{\displaystyle H}{|}}{C}}-\overset{\overset{\displaystyle H}{|}}{\underset{\underset{\displaystyle H}{|}}{C}}-\overset{\overset{\displaystyle H}{|}}{\underset{\underset{\displaystyle H}{|}}{C}}-\overset{\overset{\displaystyle H}{|}}{\underset{\underset{\displaystyle H}{|}}{C}}-\overset{\overset{\displaystyle H}{|}}{\underset{\underset{\displaystyle H}{|}}{C}}-\overset{\overset{\displaystyle H}{|}}{\underset{\underset{\displaystyle H}{|}}{C}}-H$$

The alkane family

This family of hydrocarbons is called the **alkanes**. Every chemical bond is a single bond, either between two carbon atoms along the chain (C–C) or between a carbon and a hydrogen atom (C–H). Each alkane has as many hydrogen atoms as it is possible to take – the molecules are 'full' of hydrogen. The word chemists use for 'full' here is *saturated*. Alkanes are known as **saturated** hydrocarbons.

Questions

c (i) Look at the molecular models for these hydrocarbons. If n is the number of atoms of carbon in these molecules, how many hydrogen atoms will there be? (Hint: Don't forget the ends!)
(ii) Complete the general formula for this type of hydrocarbon: $C_nH_{\underline{}}$.
d Give the formula for hydrocarbons like this with chains of:
(i) 7 carbon atoms
(ii) 22 carbon atoms
(iii) 125 carbon atoms.

Question

e What are the physical properties of carbon and hydrogen? Compare them with the physical properties of oil.

Properties of alkanes

For a liquid to boil, the molecules have to move fast enough to break free from one another. The molecules move faster when the liquid is heated. When they have enough energy to escape, the liquid boils.

Formula	Boiling point (°C)
C_5H_{12}	36
C_6H_{14}	69
C_7H_{16}	99
C_8H_{18}	
C_9H_{20}	151
$C_{10}H_{22}$	174

Questions

f From the data, which are easier to make move faster, short-chain or long-chain molecules?
g Plot a scatter graph of number of carbon atoms against boiling point and draw a line of best fit. Use this to predict the boiling point of C_8H_{18}.

Key points

- Crude oil is a mixture of different compounds. Most of these compounds are saturated hydrocarbons called alkanes.
- C_2H_6 is an example of an alkane. The general formula is C_nH_{2n+2}.
- Different sized hydrocarbon molecules have different numbers of carbon atoms joined in a chain. Short chains have low boiling points and long chains have high boiling points.

A mixture of hydrocarbons

Crude oil is a mixture of many useful hydrocarbons that need to be separated out before they can be used. This is done by a form of distillation using the physical property of boiling points.

Fractional distillation

If you heat crude oil enough it will boil. In an oil refinery, crude oil is heated strongly in a furnace so that it all boils. The gas then passes up through a tower, which is hot at the bottom but cold at the top.

> **Question**
>
> **a** Steam condenses back to water at 100°C. Paraffin boils at 200°C. At what temperature will paraffin gas condense back to liquid?

Long-chain alkanes with high boiling points will condense out quickly. Short-chain alkanes with low boiling points will have to be cooled down much more before they condense. So the different alkanes condense out at different levels of the tower. They are collected in trays and can be piped off.

This method gives a good separation, but it is not perfect. The liquid that collects at each level is still a mixture, but has a much narrower range of carbon-chain lengths, all with very similar properties. Petrol, for example, contains a range of molecules that have between 6 and 10 carbon atoms. Each liquid that is collected is called a **fraction**, so the process is called **fractional distillation**. This process can run continuously, which helps to keep the costs down.

> **Question**
>
> **b** Where will short-chain alkanes condense out, at the top or bottom of the tower?

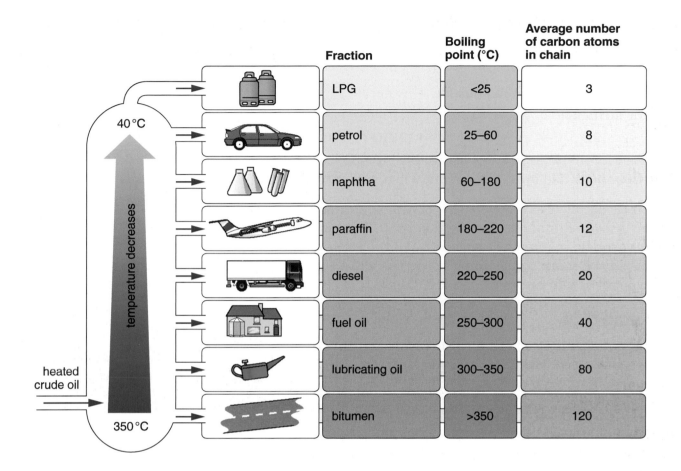

Fraction	Boiling point (°C)	Average number of carbon atoms in chain
LPG	<25	3
petrol	25–60	8
naphtha	60–180	10
paraffin	180–220	12
diesel	220–250	20
fuel oil	250–300	40
lubricating oil	300–350	80
bitumen	>350	120

40°C

temperature decreases

heated crude oil

350°C

Supply and demand

Petrol makes up about 40% of all the crude oil products sold.
Oil from different regions gives different amounts of each fraction when distilled.

Source	Petrol (%)	Paraffin (%)	Diesel (%)	'Heavy oils' (%)
Arabian	18	12	18	52
Iranian	21	13	20	46
North Sea	23	15	24	34
Demand	*39*	*11*	*30*	*20*

Questions

c *Refineries have to make enough petrol and diesel to meet the demand. To make 39 barrels of petrol they will need to distill nearly 200 barrels of crude oil. Which fractions will be left over 'unsold' if they do this?*

d *From the figures and your answer to **c**, why do you think North Sea oil is sold at a higher price than Arabian oil?*

You will find out what they do with all the leftover fractions in Section 4 of this unit.

Key points

● The physical property of boiling point can be used to separate the mixtures in crude oil by fractional distillation.

● Long-chain hydrocarbons have high boiling points and condense at the bottom of the tower. Short-chain ones condense at the top of the tower.

Carbon dioxide and global warming

Oil is mostly made from hydrocarbons. We burn oil to get energy from the reaction:

hydrocarbon + oxygen → carbon dioxide + water + **energy**

For example, for methane:

$$CH_4 + 2O_2 → CO_2 + 2H_2O$$

These waste products of burning are not 'dangerous' compounds. You make them in your body and breathe them out when you respire!

Burning fossil fuels has increased the amount of carbon dioxide in the atmosphere by over a third over the last 200 years. This may have caused **global warming**, raising the temperature of the Earth by a few degrees. This might not sound much, but it makes the weather more extreme, melts polar ice and causes the sea level to rise.

Sulfur and acid rain

Oil contains traces of sulfur. When sulfur reacts with oxygen and then water it makes sulfuric acid which causes **acid rain**.

$$S + O_2 → SO_2$$
$$2SO_2 + 2H_2O + O_2 → 2H_2SO_4$$

Forty years ago, sulfur dioxide from power stations and factories in Britain formed clouds of acid rain that caused great environmental damage both here and abroad. Acid rain from northern Britain blew all the way to Norway and killed trees and poisoned lakes there.

Today sulfur is removed from petrol and diesel before it gets to the pumps. Power stations remove sulfur dioxide from their waste gases by spraying slaked lime to neutralise the acid.

Question

a What are the waste products of burning hydrocarbons?

Question

b Explain how we think carbon dioxide causes global warming.

Questions

c What compound will you get if you use slaked lime (calcium hydroxide, $Ca(OH)_2$) to neutralise sulfuric acid?

d Complete the balanced equation for this reaction:
$Ca(OH)_2 + H_2SO_4 →$ _____ $+ 2H_2O$

Soot and global dimming

Diesel cars and lorries release billions of tiny carbon particles into the air. They are much smaller than soot from coal. You can sometimes see clouds of black smoke coming from the exhaust of a badly running lorry or bus engine.

Over the last 20 years, the sunlight falling on the Earth has been getting weaker and weaker. Scientists have taken measurements that show that energy levels reaching the Earth have dropped by 20%. They call this **global dimming**. Measurements from space show no change from the Sun itself, so scientists looked for the cause of this mysterious effect.

The tiny particles of carbon formed when oil burns get up into the atmosphere and water droplets form around them, making clouds. These clouds act like a mirror, reflecting some of the Sun's energy back out into space.

Global warming versus global dimming

So burning oil seems to be warming the Earth up with its carbon dioxide, but cooling it down with its carbon particles. The effect of global warming is kept in check by the global dimming.

Questions

f What will happen as we try to clean up the pollution caused by the carbon particles? Scientists (and politicians) will need to think very hard about this issue and be very careful about the action they take if we are to overcome this global problem.
g Your friend suggests we could stop global warming by making more diesel smoke pollution. Is she right? Explain why this might not be such a good idea.

Key points

- Most fossil fuels contain a little sulfur. This combines with oxygen to produce sulfur dioxide, which causes acid rain.
- The carbon in fuels produces carbon dioxide, which causes global warming. Some carbon particles are also released from burning fuels, which cause global dimming.
- Sulfur can be removed from petrol and the sulfur dioxide removed from waste gases to reduce pollution.

Question

e Jet engines leave 'vapour trails' of tiny ice crystals that form from their waste gases. Normally the congested skies over the USA are full of these vapour trails. After the attacks on the Twin Towers on 11 September 2001, all flights over US airspace were grounded for three days. The difference between the day and night temperatures over the USA rose markedly as the area warmed up more by day. The pattern returned to normal when air travel resumed. Suggest a reason for this.

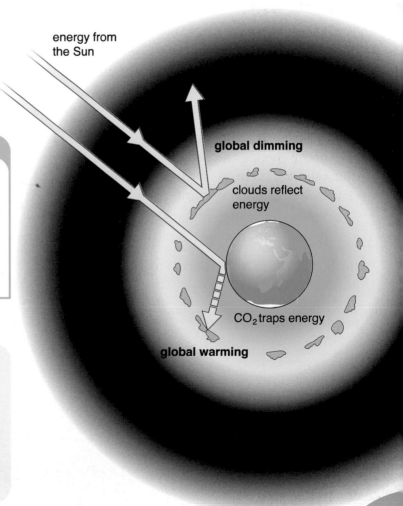

energy from the Sun

global dimming

clouds reflect energy

CO_2 traps energy

global warming

Burning fuels and global warming

Scientists agree that carbon dioxide is a **greenhouse gas** and that the level of carbon dioxide in the atmosphere is rising, at least in part due to our burning oil and other fossil fuels. There is evidence that the Earth's climate is changing as global temperatures rise. The finger of blame for global warming points to us and our use of fossil fuels. But is it really that clear-cut?

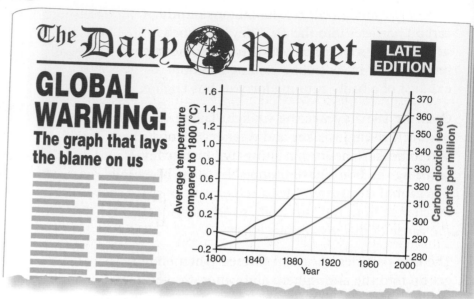

The Daily Planet — LATE EDITION

GLOBAL WARMING: The graph that lays the blame on us

▲ How carbon dioxide levels and global temperature have changed over the last 200 years.

Question

a Compare the trends shown by the two lines. Does this graph prove that carbon dioxide causes global warming?

Is it just us?

We put another 8 billion tonnes of carbon dioxide into the air every year just by burning oil. But the picture is complicated by the fact that there are other major natural sources of atmospheric carbon dioxide.

- Wildfires in Indonesia in 1997 burnt for many weeks. Over that short period it is estimated that as much as 2 billion tonnes of carbon dioxide were released into the air.
- Volcanoes release billions of tonnes of carbon dioxide into the air when they erupt.

Both of these natural sources are sporadic but at times spectacularly large. They make simple **correlation** of human action to global warming very difficult.

Questions

b Natural forests re-grow after fires. How will that affect the carbon dioxide levels in the atmosphere?

c The amount of carbon dioxide we add by burning fossil fuels is probably matched by natural sources. But every 1000 years or so you might get 100 times as much from a single 'super' volcano eruption. What would you expect this to cause?

▲ Erupting Guagua Pichincha volcano looms over Quito, Ecuador.

Global warming could be worse

Computer models of our climate predict that if all the carbon dioxide we produce stayed in the atmosphere, global warming would be much worse. But 'only' 3 billion tonnes of the gas seems to stay in the air. Let's look at where the rest goes.

Plants absorb carbon dioxide during photosynthesis and use some of it for plant growth. In old forests new growth using carbon dioxide is balanced out by dead wood that rots and produces carbon dioxide. But new forests can take up to 2.5 tonnes of carbon dioxide from the air per hectare every year!

Carbon dioxide dissolves in the oceans. Some is then taken out permanently by shellfish in their shells and ends up as limestone. Floating seaweed takes up carbon dioxide as it grows. When it dies, it sinks to the bottom and is buried in the rocks, forming the fossil fuels of the distant future.

How can we help?

The link between burning oil and global warming is not completely proven. However, we should try to limit the amount of energy we use and so reduce the amount of fossil fuels we burn.

Science to the rescue?

One possible answer to global warming is to find ways to trap the carbon dioxide before it gets into the atmosphere. We currently get our natural gas from the North Sea oilfields. The gas is trapped beneath domes of clay, deep below the seabed, but this gas will soon run out. Scientist are now trying experiments to see if carbon dioxide from power stations can be pumped back down these old wells and stored safely underground.

Key points

- Burning fossil fuels has contributed to an increase in carbon dioxide levels.
- There are natural sources of carbon dioxide, such as wildfires and volcanoes.
- Some carbon dioxide is absorbed by plants and oceans.
- We still need to try to reduce our fossil fuel use to reduce the impact on the environment.

Question

d Look at the table below. From these figures, which country looks better placed to significantly reduce its net carbon dioxide emissions by planting new forests? Explain your answer.

Source	UK	Canada
area (million hectares)	24	850
annual net carbon dioxide production (million tonnes)	550	740

Question

e The Sargasso Sea is a great mass of floating seaweed in the middle of the Atlantic Ocean. The Amazon jungle is an ancient forest. Which is better at removing carbon dioxide from the atmosphere? Explain your answer.

Question

f We have raised our own standards of living by burning oil. Are we morally justified in trying to stop developing countries from following our example?

▼ Could old oil or gas wells save us from global warming?

What's left?

Many people think we're approaching the halfway mark on our oil reserves. When will the oil run out, and what can we do about it?

Type of oil reserve	Billions of barrels of oil
oil used in the last 100 years	620
known oil reserves	1050
estimated reserves as yet undiscovered	450

Question

a From these figures you might think we have enough oil left for another 200 years. Why won't the remaining oil last anything like that long?

Hydrogen as a future fuel?

Whatever happens, oil will run out eventually, so scientists are looking for new sources of energy. Hydrogen has been used to fuel the Space Shuttle for decades, but now scientists are developing hydrogen as a fuel for buses. When this burns it just makes water, so it is completely pollution-free.

hydrogen + oxygen → water + **energy**

Some school bus fleets in California have already converted their buses to run on hydrogen.

But there are problems with hydrogen. It has to be compressed and can be quite dangerous to store if not handled properly. Also, at the moment it can only be made in large amounts by electrolysis of water using electricity.

Questions

b Hydrogen only becomes a 'renewable' fuel if the electricity used to produce it is generated by a renewable source. List three possible generating methods using renewable resources.

c At night, power stations make more electricity than they need, which is simply wasted. How could hydrogen-generating plants be used to 'store' this waste energy?

Countries with no oil

Many scientists are looking for renewable fuel resources to replace the oil as soon as possible. A vast amount of energy pours onto the Earth from the Sun every day. We can harness this for the natural process that gave us our oil reserves in the first place – photosynthesis.

Brazil has no oil of its own so took a novel approach to fuelling its cars. Instead of petrol, Brazilian motorists use alcohol. Sugar cane grows well in the hot sunshine of tropical Brazil. This sugar is fermented with yeast to produce a solution of alcohol in water. The alcohol is called **ethanol**. This is then distilled to give pure alcohol, which works well in petrol engines. Ethanol is a clean, renewable fuel. It has helped Brazil to halve its oil imports. The downside has been that large areas of rainforest have been cleared to grow the sugar cane.

▲ Sugar cane grows well in hot climates.

Questions

d Complete this equation for burning ethanol:
$2C_2H_5OH + 6O_2 \rightarrow$ _____ $+ 6H_2O$

e List the environmental advantages and disadvantages of using ethanol for cars instead of petrol. Look at the complete 'life-cycle' of the fuel, from production through to combustion.

f (i) Suggest a reason why we can't grow sugar cane in Britain.
(ii) We can grow sugar beet. What would be the advantages of making alcohol from this to run our cars?
(iii) Britain is a small country with a very large number of cars. Suggest a problem we might face if we tried to go over to alcohol instead of petrol.

Many other plants store energy in their seeds as oil – olives, sunflowers, soya beans and so on. These oils can be made to burn well in engines. This new fuel is called **biodiesel**. You will find out more about it in Section 5 of this unit.

Key points

- Oil will run out one day but scientists have already developed new fuels, such as ethanol and biodiesel.
- The use of hydrogen as a fuel is still being developed.
- Some countries are already making good use of renewable fuels.

1 a Match words **A**, **B**, **C** and **D** to the formulae in the table.

A calcium hydroxide
B carbon dioxide
C calcium oxide
D calcium carbonate

1	$CaCO_3$
2	$Ca(OH)_2$
3	CaO
4	CO_2

b Complete and balance these chemical equations.

i $CaCO_3 \rightarrow$ _____ + _____
ii $CaO +$ _____ $\rightarrow Ca(OH)_2$ *(2 marks)*

2

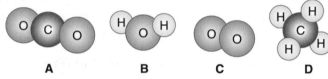

| A | B | C | D |

a Which of these diagrams shows a molecule of an element? *(1 mark)*

b Which of these diagrams show a molecule of a compound? *(1 mark)*

c i Give the chemical formulae for molecules **A–D**. *(2 marks)*

ii Give the chemical names for molecules **A–D**. *(2 marks)*

d Which subatomic particle is shared between atoms in molecules like this, to make the chemical bonds? *(1 mark)*

3

This building is made from concrete and glass.

a What are the raw materials used to make **i** cement and **ii** glass? *(2 marks)*

b What problem might you face working in a glass-fronted building like this on a sunny day? *(1 mark)*

c Suggest **one** advantage of building with concrete rather than blocks of limestone. *(1 mark)*

d List **three** environmental problems that are associated with quarrying limestone. *(3 marks)*

4 Limestone ($CaCO_3$) breaks down to quicklime (CaO) when heated. Zoe, Mumtaz and Tomi performed an experiment to see how much quicklime they could get by heating 10 g of limestone. They set their balance to zero with a crucible in place and measured out exactly 10 g of limestone. They then heated the crucible strongly, reweighing it regularly. They kept heating until the mass stopped going down. They each repeated their experiment three times. Here are their results.

	Quicklime produced from 10 g of limestone (g)			
	Expt 1	Expt 2	Expt 3	Mean
Zoe	5.6	5.7	5.5	
Mumtaz	5.65	5.66	5.64	
Tomi	5.62	5.68	6.92	

a Why does the mass as measured on the balance go down during this reaction? *(1 mark)*

b Why did they have to keep heating 'until the mass stopped going down'? *(1 mark)*

c Calculate the mean result for each of the three students. *(3 marks)*

d i Which student had been given an older, less precise balance to work with? *(1 mark)*

ii How will this have affected her results? *(1 mark)*

e Which student appears to have obtained the most reliable results? Explain your answer. *(2 marks)*

f Which student ran out of time and didn't heat her final piece of limestone long enough? Explain your answer. *(2 marks)*

g **i** The theoretical amount of quicklime produced from 10 g of calcium carbonate is 5.6 g. Whose final answer appears to be the most accurate? *(1 mark)*

ii A detailed analysis of the limestone used shows that the residue after heating is indeed slightly greater than 5.6 g. Suggest a possible reason for this. *(1 mark)*

5 Iron is made by heating iron oxide with coal in a blast furnace.

a Which element is coal mostly made of? *(1 mark)*

b Complete this word equation.

iron oxide + carbon → iron + _____ _____ *(1 mark)*

c Which reactant is oxidised in this reaction? *(1 mark)*

d Which reactant is reduced in this reaction? *(1 mark)*

e Why can't aluminium be produced from aluminium oxide by this reaction? *(1 mark)*

6 Myanmar is a poor country with underdeveloped heavy industry. Bamboo grows well in its hot climate.

In London, scaffolding is built from steel tubes that are screw-clamped together. In Myanmar, scaffolding is built from bamboo poles lashed together with natural string.

a Which do you think would be stronger, the steel or bamboo? *(1 mark)*

b What would you notice if you picked up a steel pole and a bamboo pole? *(1 mark)*

c Suggest **two** reasons why bamboo is used in Myanmar rather than steel. *(1 mark)*

d Steel scaffolding poles last longer than bamboo poles. Why is that not a problem in Myanmar? *(1 mark)*

e Broken bamboo poles are simply thrown away. Why is that not a problem? *(1 mark)*

f Large bamboo poles cost more in the UK than their steel equivalents. Suggest **two** reasons for this. *(2 marks)*

7 Look at the table:

Metal	Melting point (°C)	Strength (1 = low, 50 = very high)	Cost (£/tonne)	Density (g/cm³)
aluminium	660	1 (pure) 5 (alloyed)	18000	2.7
steel	1540	20	300	7.7
titanium	1670	10	12000	4.5
tungsten	3400	50	9000	15.3

a For each use below, suggest a suitable metal and give a reason (from the property table).
i the barrel of a Bunsen burner *(1 mark)*
ii the wing of a supersonic fighter *(1 mark)*
iii the filament in a light bulb *(1 mark)*
iv a commercial aeroplane *(1 mark)*

b Why is aluminium never used in its pure form? *(1 mark)*

c Explain in simple terms why alloys are harder and stronger than the pure metal. *(2 marks)*

8 The permitted levels of some metal ions in drinking water are:

copper	1 mg per litre
lead	0.05 mg per litre
zinc	5 mg per litre

a From these figures, which metal is most toxic, and which is least toxic? *(1 mark)*

Cattle in the fields around the river shown on the map became ill and metal poisoning was suspected. Water samples taken from the rivers at **A**, **B**, **C**, **D** and **E** were analysed and the results were:

	Concentration (mg per litre)		
	copper	lead	zinc
A	0.05	0.001	0.05
B	5.00	0.1	3.0
C	6.0	5.0	10.0
D	1.00	0.02	0.6
E	1.6	1.0	2.0

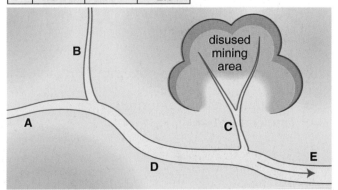

b i Which site would give safe drinking water (in terms of metal content)? *(1 mark)*

ii Which site shows the most polluted water? *(1 mark)*

iii Where do you think this pollution has come from? *(1 mark)*

c i The rivers flow from **A**, **B** and **C** to **E**. Why are the metal levels lower at **E** than at **C**? *(1 mark)*

ii From the figures, which river carries more water, the main river at **D** or the side river from **C**? By how much (to the nearest whole number)? *(1 mark)*

d i Site **A** has a full range of wildlife. How would you expect site **E** to compare with this? *(1 mark)*

ii You find water snails at site **A** but they have disappeared from the river at site **D**. Which metal do you think the snails might be sensitive to? *(1 mark)*

e The herd of cattle could only get to the river to drink at **E**. Which metal is most likely to be responsible for the poisoning? Explain your answer. *(2 marks)*

f i Scientists have suggested that the pollution problem could be tackled by throwing scrap iron into the rivers at **B** and **C**. How would this work? *(1 mark)*

ii Which metal would not be affected? *(1 mark)*

iii Which metal ion would increase in the water at **E**? Would this be a problem? *(2 marks)*

9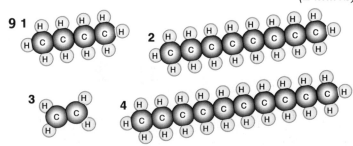

The picture shows some hydrocarbon molecules.

a Write the chemical formulae for molecules **1–4**. *(4 marks)*

b Which molecule is not an alkane? Explain your answer. *(2 marks)*

c Propane (C_3H_8) is another alkane. Complete and balance this equation for the combustion of propane. *(3 marks)*

$$C_3H_8 + \underline{\hspace{1cm}} O_2 \rightarrow \underline{\hspace{1cm}} H_2O + \underline{\hspace{1cm}}$$

d i Oil often contains sulfur as an impurity. What compound forms when this burns? *(1 mark)*

ii What problem could this cause if the sulfur was not removed from fuel oils? *(1 mark)*

10 When hydrocarbons burn they produce carbon dioxide and water. The number of molecules of each produced by burning one molecule of the hydrocarbon is shown below.

Hydrocarbon	Number of CO_2 molecules	Number of H_2O molecules	Ratio H_2O/CO_2
CH_4	1	2	
C_2H_6	2	3	
C_3H_8	3	4	
C_4H_{10}	4	5	
C_5H_{12}	5	6	
C_8H_{18}	8	9	
$C_{20}H_{42}$	20	21	

a Complete the column showing the ratio of the number of water molecules to the number of carbon dioxide molecules. *(1 mark)*

b Plot a scatter graph of the number of carbon atoms against the ratio you have just calculated. *(1 mark)*

c Draw a suitable curve and describe in words the pattern you reveal. *(2 marks)*

d What global problem is thought to be caused by carbon dioxide in the atmosphere? *(1 mark)*

e Natural gas is CH_4. Fuel oil has much longer carbon chains. From your graph, suggest a reason why oil-fired power stations are worse for the environment than natural gas-fired ones. *(1 mark)*

11 a Crude oil is split into fractions by distillation. Choose from the phrases below to label **1–4** on the diagram.

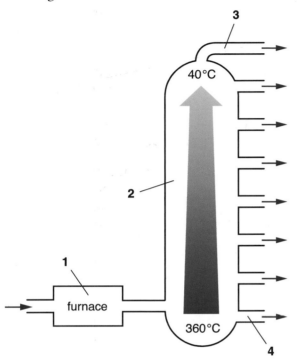

A short-chain hydrocarbons out
B crude oil vaporised
C long-chain oil out
D the vapour rises and cools *(4 marks)*

b Petrol and lubricating oil contain hydrocarbons. Petrol catches fire (ignites) easily.

Lubricating oil does not ignite easily.

Choose from the list the **two** statements that are true.
A Petrol is less volatile than lubricating oil.
B Petrol has shorter-chained hydrocarbons than lubricating oil.
C Petrol has smaller molecules than lubricating oil.
D Lubricating oil is a better fuel than petrol.
E Lubricating oil has a lower boiling point than petrol. *(2 marks)*

12 Most forms of transport work by burning fossil fuels. The table shows approximately how much carbon dioxide is produced for a 1 km journey.

bus	1000 g
large car	300 g
small car	150 g
walking	15 g

a Why is it better for the environment to drive a small car than a large one? *(1 mark)*

b Some firms run 'car pools' where people take it in turns to give each other lifts. How does this help? *(1 mark)*

c A bus can take 50 people. How much carbon dioxide is produced per person per kilometre? *(1 mark)*

d Why isn't it zero for walking? *(1 mark)*

13 Methane (natural gas) is 30 times worse than carbon dioxide as a greenhouse gas.

a Some oilfields burn off unwanted methane from the oil wells, forming carbon dioxide. Others just release the methane into the air. Which of these two options is worse for the Earth? Explain your answer. *(2 marks)*

b Trees take in carbon dioxide from the air. Cows each produce about 24 litres of methane a day from their digestive systems. In many parts of South America, the forests have been cut down to graze cattle to produce beef for hamburgers. Give **two** reasons why this is harmful to the Earth. *(2 marks)*

Crude oil is more than just a rich source of energy. In fact it has so many other uses that many people think it's a waste to just burn it. The lubricating oil and bitumen for the roads that come straight from the fractionating tower are just some of the useful products from crude oil. With a little bit of chemical know-how, many other materials can be made from oil fractions.

▼ All our plastics come from oil.

Plastic fantastic

Where would we be without plastics? From polythene shopping bags to vinyl car upholstery; from polypropene carpets to polystyrene computer housings; from a baby's bottle to false teeth and gumshields; from a Teflon non-stick coat on a frying pan to PET 'pop' bottles … Countless products are made from crude oil. Oil is a truly versatile raw material.

Smart plastics

Chemists are developing new plastics with new 'smart' properties. Some polymer gels can be made to grow or shrink. Perhaps they could be used as the muscles of future robots. New polymer transistors will make electronic devices flexible. Soon you may have a TV screen as big as your wall which rolls up neatly like a blind when you don't need it. There seems to be no limit to the possibilities of smart polymers.

▲ How about an electronic notebook that unrolls out of a pen?

Alcohol from oil

Ethanol is the alcohol in alcoholic drinks. It is made by fermenting sugary liquids like grape juice or malt mash. But vast amounts of ethanol are used in industry as an industrial solvent for varnishes, inks, paints and glues. It is also used in perfumes and aftershave. All of this industrial ethanol is made from oil.

But there are disadvantages

Plastics are easy to use in large-scale, automated industrial processes. As customers we get cheap goods, but many skilled craftsmen from carpenters to leatherworkers have been put out of work. These people need to be retrained so they can find new jobs.

Because plastics are cheap to produce, many disposable plastic products have been created, from carrier bags to cutlery, crockery and cameras. We have become a 'throw away' society. Now environmentalists are trying to encourage us to reduce, reuse or recycle what we use. Or better still, use products that aren't intended to be disposable.

Using plastics has also meant that we no longer use local, natural renewable materials such as wood and leather. Instead we rely on oil, which we will have to import when our North Sea oil runs out. Oil prices change rapidly on international markets and this can lead to economic instability. And what will we do when *all* of the oil finally runs out?

Plastics can also cause environmental problems. They are difficult and expensive to dispose of. You will consider this issue further in Spread 4.5.

Think about what you will find out in this section

What are plastics and why do they have such useful properties?

How can science help to develop fantastic new smart materials?

How are plastics made from crude oil?

How do plastics affect the environment?

How is alcohol made from oil?

Oil fractions and demand

The fractional distillation of crude oil makes many useful products. Unfortunately the demand for these does not match the supply from the refinery.

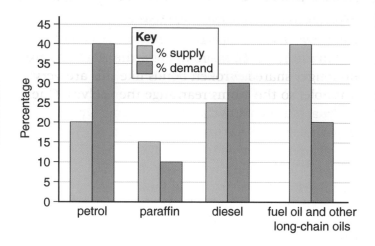

Question

a How does the supply of the different fractions compare with the demand?

To keep up with the demand for petrol by distillation alone, a refinery would end up with vast amounts of 'unwanted' long-chain oils. For every barrel of petrol it sold, there would be another three barrels of long-chain oil it couldn't sell.

Industrial cracking

Fortunately it is fairly easy to 'chop up' long-chain hydrocarbons. If they are heated, they can be broken up by a thermal decomposition reaction. This is called **cracking**.

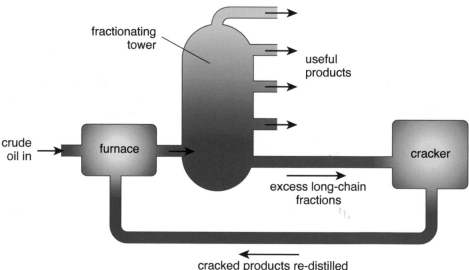

The cracking process takes place in the oil refinery.

- First the unwanted long-chain fractions are boiled.
- The vapour is then passed over a hot catalyst (this makes the reaction work faster).
- The long-chain molecules break up into smaller pieces.
- The shorter-chain products are then passed back through the fractionating tower to separate them.

The extra petrol needed is produced in this way, as well as other important chemicals used to make ethanol and plastics.

Question

b Suggest why this cracking takes place at the oil refinery, rather than in a separate plant elsewhere.

More about cracking and its products

Alkanes make their chemical bonds by sharing electrons. When an alkane is cracked, a carbon–carbon bond (C–C) is broken and the electrons are no longer shared. Broken bonds like this are very unstable, so the atoms rearrange themselves. One hydrogen atom from the smaller piece joins the larger piece to make another alkane.

This leaves the smaller piece short of two hydrogen atoms – and with two broken bonds. These bonds snap together, forming a carbon–carbon double bond (C=C). Hydrocarbons with double bonds like this are called **alkenes**. The simplest alkene, with just two carbon atoms, is called ethene.

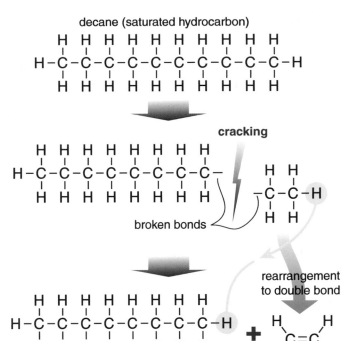

decane (saturated hydrocarbon)

cracking

broken bonds

rearrangement to double bond

octane (saturated)

ethene (unsaturated)

Question

c The diagrams show an alkane called decane being cracked to give an alkane called octane and an alkene called ethene. Write out the reaction as both a word equation and a balanced chemical equation. (Hint: You will need to count the atoms.)

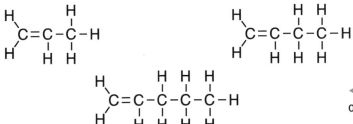

◀ Some other members of the alkene family.

Question

d What are the formulae of the alkenes shown here? What is the general rule for the formula of an alkene?

Alkenes have two hydrogen atoms less than alkane with the same chain length. So alkenes do not have the full number of hydrogen atoms. Alkenes are called **unsaturated** hydrocarbons.

Ethene is a hydrocarbon and can be burnt. In the past, because petrol was the product needed from the process, ethene was seen as a waste product and simply burnt off. Now we have lots of uses for ethene. It is used as a raw material to make a range of things from ethanol to plastic.

Question

e For every 170 kg of decane cracked in this way you could in theory get 142 kg of petrol (octane) and 28 kg of ethene. Suggest a reason why you might not actually get that much of either. What might you get instead?

Key points

- Hydrocarbons can be cracked by thermal decomposition using a catalyst to produce smaller, useful molecules.
- The products of cracking are unsaturated alkenes with the general formula C_nH_{2n}. An example is ethene, which has the formula C_2H_4.
- Some of the fractions from cracking give petrol, which is much in demand.

Traditional ethanol

Ethanol is the alcohol in wines, beers and spirits. People have been making these drinks for thousands of years. Fermenting grape juice or malt mash makes a weak solution of ethanol in water. Distilling it gives a stronger alcoholic drink with more alcohol such as whisky or brandy.

▶ Wine and beer have been around for thousands of years.

Product	Source	How it's made	Proportion of ethanol
beer	barley mash	fermentation	5%
wine	grape juice	fermentation	11%
whisky	barley mash	fermentation and distillation	40%

Question

a Explain why there is more alcohol in whisky than in beer or wine.

Ethanol in industry

Pure (100%) ethanol is also a very important industrial chemical. It is used as a solvent in glues, varnishes, inks and paints. It evaporates quickly so it is also used as a 'carrier' for perfumes and aftershave.

For large-scale industrial use, pure ethanol has to be made as cheaply as possible. Alcoholic drinks are relatively expensive – partly due to the high tax paid on them. Almost pure ethanol can be made in the same way as whisky or brandy, but the fermentation and distillation process is slow and relatively costly, even without the tax the government adds. A new source was needed and oil provided the answer.

Question

b Suggest two reasons why vodka costs so much more than methylated spirits.

85% ethanol   40% ethanol

▲ Methylated spirits is mixed with poisonous methanol, a dye and a bitter chemical to stop people drinking it.

Ethanol from waste

The cracking process that produces more petrol from oil also makes ethene as a waste product. Chemists discovered that they could react ethene with steam to give ethanol. The carbon=carbon double bond in ethene is relatively weak. With a little heat energy and a suitable catalyst the double bond breaks. The loose 'bond arms' then grab hold of passing water molecules from the steam to make ethanol.

This process is called **hydrolysis**, which just means 'adding water'. This is an addition reaction, because new atoms are added to the compound.

ethene + water $\rightarrow$ ethanol

C_2H_4 + H_2O $\rightarrow$ C_2H_5OH

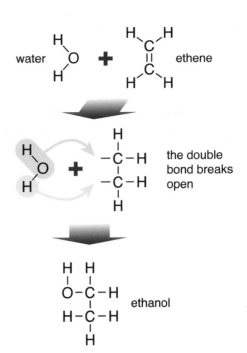

water / ethene

the double bond breaks open

ethanol

Question

c What compound would you get if you made ethene react with hydrogen (H_2)?

So which method is best?

It depends what you want the ethanol for – and whether you are thinking about now or the future.

	Fermentation	From ethene
type of process	This is a batch process. Fixed amounts of sugary solution are fermented at a time. The weak solution of alcohol is then distilled.	This is a continuous, automated process. Ethene and steam are constantly fed over the heated catalyst.
speed of production	Fermentation is a slow process that takes days or weeks.	The reaction takes place as fast as the reactants can be pumped in.
quantity of product	Limited by the size and number of fermentation vats.	In a large refinery the amount of alcohol produced can be tailored to meet demand.
quality of product	The proportion of alcohol may be low, but the impurities give it the flavour.	Pure ethanol is produced with ease. Impurities may be added later to stop people drinking it.
the raw material	Sugar comes from photosynthesis. Plants are a renewable resource.	Ethene is produced from fossil fuels. These are non-renewable.
cost	The use of batch processing and the long timescale makes this relatively expensive.	The continuous processing makes this relatively cheap – but that could change if oil prices go up.

Questions

d Breweries need to employ more people than refineries. Why do you think that is?

e What are the advantages and disadvantages of these two methods of making ethanol? Draw up a table.

Key points

- Ethene can be reacted with steam using a catalyst to produce an industrial alcohol, ethanol.
- The industrial method of producing ethanol relies on ethene, which is a by-product of cracking long-chain hydrocarbons from non-renewable oil.
- The fermentation method of producing ethanol relies on renewable plant resources.

Polymerisation

The small alkene molecules made by cracking have a huge range of astounding uses as plastics. These small molecules, or **monomers**, can be made to 'pop' together like beads to make very long chains called **polymers**. There may be thousands of monomers popped together in a polymer chain. The process is called **polymerisation**. The long chains stick together like a tangled mass of spaghetti. It is this structure that gives polymers their useful properties.

Polymers are plastics. They are very easy to shape and are great for making objects from mobile phone cases to sink units. Plastics are very versatile and are used from the 'cheap and cheerful' end of the market to high-tech luxury.

> **Question**
>
> **a** List five objects made from plastic in the room around you. Suggest why plastic was used in each case.

It's that double bond again...

In this type of polymerisation, the double bonds in the monomers break open. The 'loose' bonds from neighbouring monomers then join up to form a long polymer chain. This is called **addition polymerisation** as the chains are simply added together.

Poly(ethene)

Polymers are named by putting 'poly' in front of the name of the small molecule. The simplest molecule is called ethene, so the polymer made from this is called poly(ethene). We often shorten this to polythene.

Poly(ethene) is a waxy solid that is very easy to shape. We can mould into bottles that can be used for drinks, or even for dangerous chemicals such as bleach or acid. It can also be rolled into thin but tough, flexible and waterproof sheets. These are ideal as a food wrap or for plastic bags.

> **Question**
>
> **b** Why are polythene bags preferred to paper ones by most people? Suggest one disadvantage of using polythene bags.

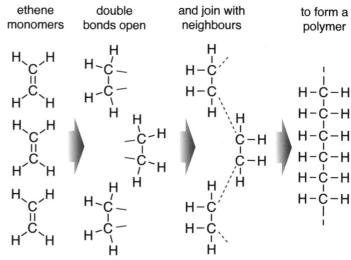

ethene monomers	double bonds open	and join with neighbours	to form a polymer

Cracking makes small ethene molecules.

These can be made to pop together to form poly(ethene).

The chains stack up like molecular spaghetti.

Poly(propene)

Poly(propene) is made from the alkene called propene (C_3H_6). It is a tougher plastic than poly(ethene) and it's not so flexible. It is used to make hard-wearing things such as bowls, crates or even school chairs. It is also made into fibres for carpets and ropes. Poly(propene) can be made brightly coloured so it is often used for children's toys.

Question

c *Poly(ethene) chains are smooth and can slide over one another. Poly(propene) chains are 'knobbly' and can't slide easily. What property does this give poly(propene)?*

The polymer for the job

Different polymers have different properties that make them suitable for different jobs.

Polymer	Flexibility	Toughness	Relative cost	Other property
poly(ethene)	high	low	low	
poly(propene)	medium	high	medium	easy to colour
poly(styrene)	low (brittle)	brittle	medium	can be precisely moulded
poly(chloroethene) (PVC)	high	high	high	resistant to corrosion a very good electrical insulator
poly(ethane terephthalate) (PET)	high	medium	medium	transparent easy to 'blow mould'
poly(tetrafluoroethene) (PTFE)	high	medium	high	low friction – 'non-stick'
poly(methyl methacrylate) (acrylic)	low	medium	high	highly transparent

Question

d *Which polymer would you use for these products and why? (i) Insulation on an electrical cable; (ii) the case for a stereo system; (iii) the coating on a frying pan; (iv) a 'shatter-proof' window.*

Smart polymers

Scientists are continually developing new polymers. Many of these new polymers seem to work well in the human body. New dental polymers are being made to replace old-fashioned metal fillings, for example. But some of these new polymers have unusual – 'smart' – properties.

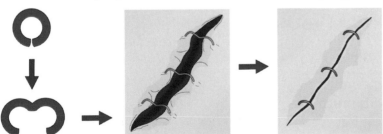

Question

e *Shape memory polymers can be used to make moulds for casting plaster or concrete. You could make the mould by warming the polymer, stretching it into shape and then cooling it.*
(i) Explain how this material could be used over and over again for different moulds.
(ii) Even if these special polymers cost three times as much as the traditionally used latex rubber, why might they still end up saving money?

Shape memory polymers can stretch like rubber when warmed but will 'stick' in their new shape if cooled. When they are warmed again they revert to their original shape. Large wounds can now be closed using special 'shape memory' polymer clips. These tight rings can be stretched out, cooled and fitted loosely into place. As they warm up, they pull back to their original shape, gently and securely closing the wound.

Key points

- Alkenes monomers such as ethene and propene can be used to make polymers such as polyethene and polypropene in a process called polymerisation.
- Polymers have many uses depending on their properties.
- Scientists are now developing 'smart polymers' with strange but useful properties.

From PVA to hydrogels

Slime is wonderful stuff, isn't it? You can mix your own in the kitchen sink. But what is it? And how can you change it from soft dripping slime to the bouncing variety?

Simple slime is made from PVA. PVA is short for poly(vinyl alcohol) – the old name for poly(ethanol). On its own it is not particularly strong, but it mixes well with water, making PVA glue. As the water dries out, the polymer chains stick to each other and whatever it is they are sticking together.

But PVA glue is not a simple mixture. The water molecules get between the polymer chains and are held weakly in place. If PVA reacts with a chemical such as borax, weak chemical bonds form 'cross-links' between the polymer chains, making a loose grid. Water molecules can get into this grid, but are held more tightly than in plain PVA. This gives slime its properties. Slime with few cross-link bonds is very runny. Slime with more cross-link bonds is much more viscous.

The open structure and trapped water makes slime soft and easy to pull apart, breaking the cross-link bonds between the chains. When two pieces of slime are pushed back together, the cross-link bonds re-form and the slime forms a single piece again.

Materials like this are called **gels**. You can make weak, low viscosity gels like slime by having lots of water and less cross-linked polymer. Or you can make strong, high viscosity, bouncy gels with more polymer and more cross-links.

Using different polymers, you can make gels with much stronger cross-links. These can hold their shape – but they can also trap lots of water. These **hydrogels** are used to make soft contact lenses.

Other hydrogels can be used in hospitals to cover wounds. They keep the wounds safe from infection, but allow small molecules like oxygen and water to move through, allowing the flesh beneath to 'breathe'.

PVA: polymer chains – no cross-links

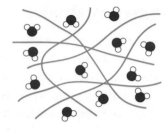

PVA glue: polymer chains – no cross-links, water molecules held loosely between chains

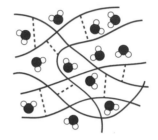

slime: polymer chains – weak cross-links, water molecules trapped loosely between chains

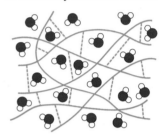

hydrogel: polymer chains – strong cross-links, water molecules trapped between chains

Question

a Explain how the number of cross-link bonds affects the physical properties of slime.

Question

b Which would a hydrogel work best on – a graze or a large open wound?

▼ Soft contact lenses are made from hydrogels.

Keeping the water out

Hydrogels keep water in – but other polymers are used to keep the water out. PTFE (polytetrafluoroethene) is the polymer used to make non-stick frying pans. A slightly different version of PTFE is used to make clothing waterproof. A thin layer of the tangled mass of polymer strands has millions of tiny pores to every square centimetre. When you sweat, the water molecules in the water vapour you give off can escape through these pores. The layer 'breathes' like cotton. But rainwater falls in droplets containing billions of tightly packed water molecules. These stick together and the whole droplet is much too big to get through the pores. So the layer is as waterproof as a polythene sheet!

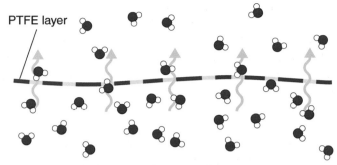

water vapour from sweat can escape

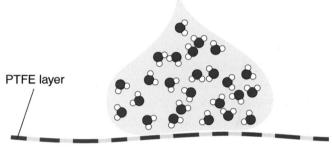

water in raindrops can't get in

Question

c What are the disadvantages of wearing polythene clothes?

Futuristic applications

Scientists have now made plastics that conduct electricity. Going further, they have developed plastics that can be used to make transistors and microchips. These plastics partially conduct electricity in a similar way to silicon. Normal microchips are very expensive to make. You have to grow silicon crystals in carefully controlled conditions. If any dust gets in, it's all ruined! Plastic microchips could be made by simply printing them using special inks on special plastic sheets. All you need is an industrial-sized version of the bubble jet printers we use at home.

Question

d What advantages will 'plastic' computer chips have over the silicon ones we use now?

In the future, plastics might dominate all aspects of our lives. Athletes will have computers built into their clothing to monitor their performance. You might have computers built into your clothing that monitor your health!

Question

e Silicon is made from sand, plastic is made from oil. What are some possible disadvantages of changing to plastics for our microchips and computers?

▲ Smart shirts could easily write backwards for mirror viewing!

Key points

- The properties of polymers depend on what they are made from and how they are produced.
- Polymers have many useful applications and new uses are being developed all the time.
- Hydrogels are useful new polymers.

Great plastics

Plastics are everywhere today. Much of our food now comes pre-packed in plastic. At home, our plastic-cased electronic equipment comes packed in plastic foam to keep it safe, and we put our rubbish into plastic bin bags before we throw it all out.

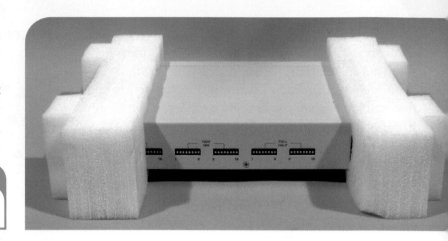

> ### Question
>
> **a** *Suggest what materials were used for packaging before plastics. What were their disadvantages?*

The down side

Disposing of plastics is a problem. The plastic age is also the over-packaged, throw away age. Over a million tonnes of plastic packaging are produced per year in the UK. In the worst case, this plastic ends up as litter – an ugly reminder of our waste on our streets, in the countryside and on the beach.

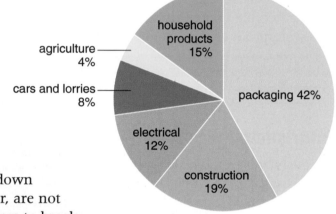

Biodegrade, tax or ban?

We throw paper away too, but this natural material rots down rapidly in the soil – it is **biodegradable**. Plastics, however, are not biodegradable, and may take tens or even hundreds of years to break down. Most plastic rubbish ends up in the dustbin and then on the local tip, which is usually a landfill site – perhaps an old quarry. The problem is that these are filling up fast.

> ### Question
>
> **b** *Using the pie chart, explain why a reduction in the amount of plastic packaging we use would make a significant contribution to the overall amount of plastic waste.*

Scientists are working on the development of smart plastics that are biodegradable. Some of these are based on natural polymers such as starch, which break down easily without leaving toxic waste. They would be more expensive than oil-based plastics, but it would be worth the extra cost to stop damage to the environment. And the price would soon come down if their use increased.

In Ireland they tax plastic bags. In New York State in the USA some counties have banned plastic bags completely. Food shops have to use paper bags instead.

Recycling and energy use?

Plastics are made from non-renewable oil. One way to use less oil is to recycle wherever possible. Most plastics have stamps that say what type they are. Many countries have set up separate recycling bins for different plastics, so that people can separate their waste at source. Maybe it should be compulsory.

Recycling uses a lot of energy, often more than it would take to make new plastic. Plastic waste contains as much stored energy as fossil fuels. Sweden already recovers 52% of the energy from plastics by burning plastic waste as fuel in power stations. Think how much oil you could save!

A detailed cost-benefit analysis shows that, overall, plastics use less energy and cause less environmental damage than traditional materials. For example, glass bottles are made from sand, but this has to be melted first. It takes 230 kg of oil to make 1000 glass bottles. You could make 1000 plastic bottles from just 100 kg of oil.

Different countries, different ways

The graph shows how different countries are dealing with plastic waste.

Amount of waste produced per person, 1995–2003

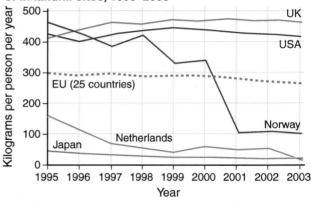

Amount of municipal waste per person disposed of in landfill sites, 1995–2003

Questions

c How does Norway compare with the UK in terms of:
  (i) the amount of waste produced?
  (ii) how they dispose of that waste?
d How has this pattern in Norway and the UK changed over time?
e Norway has banned the dumping of biodegradable materials in landfills. Do you think this accounts for all the change in their graph?
f Suggest other ways in which Norway might have reduced the scale of the problem.
g In the Netherlands, campaigns have encouraged people to reuse the plastic carrier bags they get from the supermarket instead of using new bags every time and then throwing them away. Does the waste graph suggest this has been effective?
h Japan and the USA both have advanced high-tech societies. What difference do their waste graphs suggest about these two cultures?
i Compare the proportion of waste (as an approximate fraction) that ends up in landfill for Japan and the USA. Suggest a reason for this difference.
j The EU has put a tax on landfill sites. How do you think this might affect the way local councils deal with their waste? Is it helpful?
k Copy the table below. From the information on this spread, list the social, economic and environmental impacts of each action.

Actions	Social impact	Economic impact	Environmental impact
using paper instead of plastic			
using biodegradable plastic			
using plastic instead of glass			
sorting and recycling plastic			
burning plastic for energy			
burning plastic rubbish in incinerators			

Key points

● Many polymers are not biodegradable, which causes a problem with waste disposal.
● The disposal or recycling of polymers costs money and can have social environmental and economic impacts.

Plants and oil

All plants make glucose by photosynthesis, but some plants change this glucose to oil. Oil is a more concentrated store of chemical energy than glucose. Many plants make oil in their seeds, to provide energy for their growing seedlings. We can eat these seeds and use the energy directly. Or we can take the oil from them to use in other ways, for example as fuels.

Extracting the oil

Olives have been used to make oil for thousands of years all around the Mediterranean. Traditionally, the olives were crushed by giant stone wheels rolling over them, or squashed in large, hand-turned presses. Now these are replaced by industrial hydraulic presses.

The oil is squeezed out of the crushed pulp, runs out and is collected. The oil is then separated out from any water or other impurities. Even with modern methods olive oil is expensive, as olives are difficult to harvest and contain less than 20% oil.

Oil is sometimes removed from plants by distillation, but this can alter the flavour and smell of edible oils. Some plants like lavender produce small amounts of scented oils. These can be removed by distillation with water, which works at a lower temperature.

Oilseed crops

For cooking olives have been replaced by oilseed crops that can be harvested and processed cheaply and easily. In Southern Europe, sweetcorn and sunflowers grow well. In cooler Britain, the favourite oilseed crop is oilseed rape, as it grows well in our climate and the seeds contain up to 50% oil by mass.

This close relative of cabbage with its bright yellow flowers has changed the colour of the English countryside in summer over the last 20 years. Rapeseed oil is produced cheaply by industrial-scale processes. It is also a 'healthy' oil in your diet (see Spread 5.1). But the seeds are scattered easily and you can now see this plant growing along roadsides and invading fields. There is a danger that it will push out native wild flowers.

Vegetable transport fuel?

Oils are fuels for our bodies. Vegetable oils can also be made into biodiesel as a fuel for cars and lorries. Biodiesel itself is non-toxic to plants and animals and breaks down easily in the natural environment if it is spilt. Its exhaust contains fewer pollutants than ordinary diesel. It does give off carbon dioxide when it burns, but the plants that it is made from took in carbon dioxide when they grew. So it is a renewable fuel that does not cause global warming.

This could be the renewable fuel of the future, but you would need to grow a lot of oilseed. You can get about 1000 litres of oil from a $4000\,m^2$ field of oilseed rape. That's enough to run an average car for a year. But there are 24 million cars in Britain. You would need to turn $96\,000\,000\,000\,m^2$ of land over to oilseed farming if you wanted to replace all the petrol and diesel. That's almost half the total land area of Britain, so we could never become self-sufficient here. Biodiesel could still make a significant contribution to our energy needs in the future.

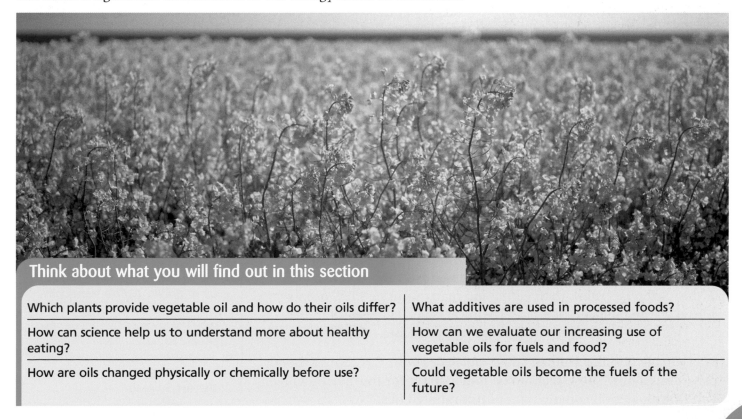

Think about what you will find out in this section

Which plants provide vegetable oil and how do their oils differ?	What additives are used in processed foods?
How can science help us to understand more about healthy eating?	How can we evaluate our increasing use of vegetable oils for fuels and food?
How are oils changed physically or chemically before use?	Could vegetable oils become the fuels of the future?

Healthy or not?

We seem to get conflicting messages about fats or oils in our food. But if you understand the science it helps.

▲ Oily food is bad for you!

▲ Oily food is good for you!

Oils for energy

You must have some oil in your diet for your body to work properly. For example, you get vitamin A from oily foods. Oils and fats are primarily energy foods; fats are just solid oil. You need energy for life and the more active you are, the more energy you use. Oils are a more concentrated energy food than carbohydrates. Every gram of oil provides 39 kilojoules (kJ) of energy – double that from carbohydrates. An averagely active adult needs about 12 000 kJ of energy every day, which is the equivalent of 300 g of oil.

What sort of oil?

Vegetable oils have long carbon-chain molecules but they are not just simple hydrocarbons. At their simplest, all the C–C bonds are single bonds. These are called saturated oils. Animal fats are also 'saturated' like this. Scientists think than eating too many saturated oils or fats can lead to the build-up of a fatty chemical called cholesterol in the body. This can block arteries and cause heart disease.

But vegetable oils also contain some molecules with C=C double bonds – unsaturated oils. These seem to be much better for our health. Olive oil and sunflower oil contain a lot of unsaturated oils.

Questions

a The main oil available in Mediterranean countries is olive oil. In Britain, people traditionally cooked with butter or lard. Why would you expect Mediterranean people to have a healthier diet?

b Suggest how people have changed the way they use oil for cooking in the last 50 years.

Oils for health

The main health problem associated with eating large amounts saturated fat such as lots of fried food is the increased risk of heart attack or stroke. This appeared clear-cut at first, but scientists studying the diets of the Inuit of Canada found that their results did not fit the simple pattern.

Further study of the Inuit diet showed that their health depended on the fact the animal oil they ate was from fish, and fish has a particular type of oil in it.

People studied	Main oil source in diet	Risk from heart disease
Greeks	olive oil	low
Scots	lard, beef suet or butter	very high
Canadian Inuit	whale or seal blubber and fish oil	very low

Your body can make most of the oils it needs if necessary, but there are two types of unsaturated oil you must get from your diet because your body can't make them. They are called omega-3 and omega-6 oils. Oily fish and seafood are rich sources of these oils. The large amounts of these oils in the Inuit diet counteracts the saturated fats they get from whale and seal meat. It helps to 'thin' the blood and stop the build-up of cholesterol. Omega-3 oils are also good for your brain and help your memory work efficiently.

Vegetarians need omega oils too. Fortunately some seed oils contain these special oils. Flaxseed oil has just the right balance of omega-3 and omega-6 oils for the human body. Many people now buy flaxseed oil capsules from health food shops as a dietary supplement.

How can you tell which is which?

Bromine water is an orange–brown solution containing bromine (Br_2) molecules. It is not reactive enough to react with saturated oils. But the C=C double bonds in unsaturated oils make them more reactive than saturated oils. The double bonds break open and react with bromine and an addition reaction occurs. The new compounds that form are colourless, so the bromine water loses its colour. This reaction is used as a test for unsaturated oils. A weak, brown iodine solution can be used in a similar way.

Question

e Which is more likely to decolorise iodine solution, olive oil or lard? What is happening to cause this colour change?

Question

c The Greek and Scottish people studied followed the pattern seen elsewhere. In what way were the Inuit results anomalous?

Question

d Why might it be a good idea to eat mackerel or salmon when revising for an exam?

▲ A field of flax.

Key points

- Vegetable oils are important foods because they provide lots of energy and nutrients.
- Different types of oils have different effects on our health.
- Unsaturated vegetable oils with carbon=carbon double bonds can be detected using bromine or iodine.

Ready to eat

Fresh food may be best for our health, but in our busy modern world processed foods are certainly quicker. What happens to food when it is processed?

Changing vegetable fat

Fats are just oils that are solid at room temperature – they have higher melting points. Most natural fats come from animals, for example lard from pigs, suet from cows or butter from milk. Fats are better than oils for making cakes or pastry, but they are linked to heart disease. And they are also 'not suitable for vegetarians …'.

Saturated oils have higher melting points than unsaturated oils. Some vegetable products like coconut oil are high in saturated oils and do turn solid when cold. But these also have a strong flavour that isn't suitable for all food. Scientists have found a way to change cheap unsaturated oils into saturated fats for cooking.

The oil is warmed to 60 °C and hydrogen gas is bubbled through it. The double bond snaps open and 'grabs hold' of a passing hydrogen molecule to become saturated. Nickel is used as a catalyst to speed up the reaction. The process is called **hydrogenation**. Hydrogenated oils are used as fats for margarine, to replace butter. They are also used in cakes and pastries – and in chocolate!

▲ Fresh cooked spaghetti and calamari, or …

▼ a TV dinner … which is healthier?

hydrogen

H–H

+

---C–C=C–C---
    | |
    H H

unsaturated oil

H–H

---C–C–C–C---
    | |
    H H

the double bond breaks open

   H H
   | |
---C–C–C–C---
   | |
   H H

saturated oil

Question

a *Expensive chocolate is made with the unsaturated fats found naturally in cocoa butter. Why do you think this is often partly replaced by hydrogenated vegetable oil?*

Additives in our food

Processed foods can lose some of their flavour and natural colour. They also need to last a long time on supermarket shelves. Processed foods have additives put in to overcome these problems. Some of these are 'natural' chemicals that have been used for thousands of years. Others have been created by scientists to do specific jobs.

- *Flavour enhancers* – We all like a little salt and maybe vinegar on our chips, or sugar in our baked beans. Monosodium glutamate is used in Chinese cooking, and we find it in lots of foods. It may sound as if it comes from a chemistry lab, but it is actually made from fermented soya beans!
- *Colour enhancers* – Processing dulls the colours of foods like peas and tomatoes. Some products use food colours to improve the appearance of the food. Some are natural, like the 'carotenes' extracted from tomatoes and carrots. Other are synthetic – made in the laboratory – such as yellow tartrazine and carmine red.

- *Preservatives* – Vinegar for pickles and sugar for jam have been around a long time; chemicals such as benzoic acid or sulfur dioxide are more recent additions. These preservatives all stop microbes growing in the food. Another group, called antioxidants, stop fats and oils going rancid and tasting bad.
- *Vitamins and minerals* – These are sometimes added to replace those lost during processing. Others are added to improve the product. Calcium is sometimes added to milk products, while ascorbic acid (vitamin C) is added to many fruit-flavoured drinks.

Are all these chemicals good for us?

Used wisely, some chemical additives help us make the most of our food. But there are some potential problems.
- Some people might be allergic to some additives.
- Additives might be used to disguise substandard ingredients.

Additives have all been tested, but there are concerns that long-term consumption of some additives could cause health risks. Many people believe that some of the bright colours used in sweets and drinks may cause children to lose concentration and become hyperactive.

Question

b Look back at the types of additives and give four reasons why they can be helpful.

How can we tell what's in our food?

By law all processed food products have to list their ingredients. As some of these have long 'chemical' names, the government has produced a list of permitted additives and given them 'E-numbers'. For example, E100–E199 are colours, E200–E299 are preservatives and E300–E321 are antioxidants. Some people reject any food with E-numbers in them as they think additives must all be nasty, harmful chemicals. But some of them are perfectly 'natural'. E300 is vitamin C while E260 is acetic acid, or vinegar.

Questions

c Explain to a friend who is against foods containing E-numbers why some additives can be beneficial.
d Suggest as many reasons as you can why people eat processed foods rather than fresh food.

Key points

- Vegetable oils can be hardened by hydrogenation so they can be used in spreads, cakes and pastries.
- Processed foods contain additives to improve taste and appearance, and to preserve them, but there are also disadvantages of using additives.
- Permitted additives are classified using the E-number system.

Just paint?

You may have used a can of white emulsion to paint your bedroom. But you use emulsions more often than you might think.

So what is an emulsion?

Oil and water do not mix. If you shake water with a small amount of oil, tiny droplets of oil will spread throughout the water. An **emulsion** is a mixture of tiny droplets of oil in water or water in oil. If you let the mixture stand, the oil and water will separate out. This happens quickly with an oil and vinegar salad dressing.

Question

a Why does the oil rise to the surface?

The trick with emulsions is to stop them from separating out. This happens naturally in emulsions such as milk, cream and butter. Milk is an emulsion of a few per cent of tiny droplets of butterfat in a watery liquid. Cream is similar but has more fat and less water. In butter, the emulsion is the other way around – with just a few per cent of milky water as droplets within the fat.

Even milk and cream separate eventually, however. The oil in milk slowly rises to the surface where it can be skimmed off – as cream. Butter is made by churning cream, which concentrates the fat even more. But some water is left. Once you melt butter, however, it separates completely and you can't get the butter texture back. The resulting butterfat is called ghee.

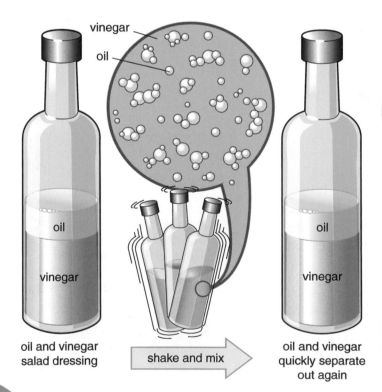

vinegar
oil

oil

vinegar

oil and vinegar salad dressing

shake and mix

oil

vinegar

oil and vinegar quickly separate out again

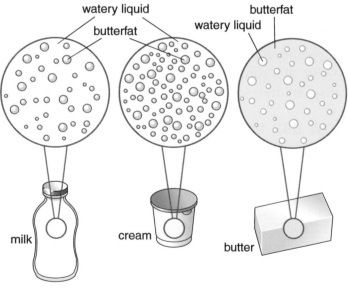

watery liquid
butterfat
butterfat
watery liquid

milk

cream

butter

Question

b Why do people on diets drink skimmed milk rather than ordinary milk?

Preventing separation

Mayonnaise is made from oil and vinegar but it doesn't separate out because it also contains a little egg yolk. Chemicals in the egg yolk stop the tiny droplets of oil from joining together when they collide, so the emulsion does not separate out. Chemicals that act like this are called **emulsifiers**. They are so important in the food industry that they have a whole section of E-numbers to themselves (E322–499).

Emulsions are very versatile, but even with the best emulsifiers they will separate out in time. This limits the shelf-life of products containing emulsions.

Why are emulsions so useful anyway?

Emulsions are thicker than oil or water. The thickness depends on the amount of oil and water, and how it is arranged. How we use them depends on their special properties.

Emulsion paints were first introduced to make paint more 'user-friendly' for the DIY market. They had less oil in them so they smelt less and dried quicker. You could even wash your brushes in soapy water instead of needing white spirit. But making an emulsion also changed the texture. Emulsions were less runny than oil-based paints and with a little tweaking could even be made 'non-drip'. They stay in place on the brush and do not run on walls.

This thicker texture is also important in food emulsions. Emulsions may be made to coat other foods, to keep particular shapes or simply to give a creamy, even texture. Mayonnaise is much thicker (more viscous) than salad dressing. It doesn't pour, it dollops – and sticks nicely to your chips.

Cream is less runny than milk; double cream is even 'thicker' as it has more fat. But whisk in some air bubbles and you get stiff whipped cream that sticks to your strawberries or can be piped into pretty shapes on your trifle. Ice cream is another water/fat emulsion that has the added complexity of tiny air bubbles – and ice crystals.

Questions

c Some emulsifiers give the droplets a positive electrical charge. Like charges repel one another. How does this stop the droplets joining together and the emulsion from separating?

d Jars of very old and stale mayonnaise often have an oil layer in them. Explain what has happened.

Question

e Presentation is very important when selling food. Give examples where the careful use of emulsions improves the presentation of processed food.

The way emulsions change the physical properties of the ingredients is vitally important to the food processing industry. Creamed cake mixture is an emulsion of water in fat. It is soft enough to be poured into a tin, but firm enough to keep the ingredients well mixed until it is cooked. Soft margarines are also water-in-oil emulsions. The water in the margarine makes it soft enough to be spread straight from the fridge.

Key points

- Emulsion are mixtures of oil and water, which can be made to stay together using emulsifiers.
- Emulsions have different uses depending on their properties. Examples are milk, butter, mayonnaise and paint.

Sudan 1

Fresh chillies can be bright red. When they are dried and processed to make chilli powder the colour usually darkens. Food colours may be added, to make them look fresher and more appealing.

Harmless colours are fine but sometimes dyes are used that should not be put in food. Some of these could cause long-term health problems such as cancer. Sudan 1 is one of the dyes that is not allowed in food. In 2005, Sudan 1 was found in processed food. Around 400 products had to be withdrawn from supermarket shelves.

> **Question**
>
> **a** List five types of food that might have contained chilli.

How can we tell what's in our food?

Foods are mixtures. We have to separate out the chemicals in the food before we can identify them. One of the simplest ways to separate out artificial food colours is to use **chromatography**. You put a spot of food colour onto a piece of filter paper and let water soak up through it. The different colours move up the paper with the water at different rates. You can then compare the dyes in the food with a pure sample of the dye you are investigating.

Chemical analysis

Simple chromatography might be fine if you just want to check what to avoid in your food. But before withdrawing millions of pounds worth of food from the supermarket shelves you need to be very sure of your facts. Supermarkets use analytical laboratories which can pinpoint chemicals accurately even at very low concentrations.

To test for Sudan 1, scientists used a special kind of chromatography to separate out the different chemicals in chilli powder. These were then fed into a machine called a mass spectrometer, which gives a printout that acts like a chemical fingerprint. The 'fingerprint' was then matched against a database of known dyes.

Of course, this system is only as good as its database. Just as a fingerprint found at the scene of a crime is of no use unless the police can match it to a known criminal, this system is only effective for 'known' chemicals. Laboratories in many universities across the world are constantly analysing new chemicals and adding their results to the global database. But some companies will only release their data if you pay for it.

From database for comparison

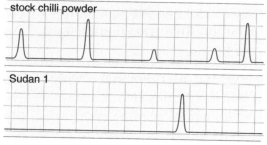

Chilli powders being tested

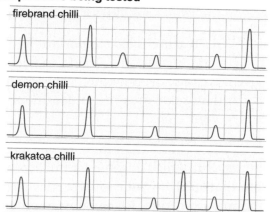

> **Question**
>
> **b** Which of these chilli powders contains Sudan 1?

> **Question**
>
> **c** Should all chemical data like this be freely available? Suggest reasons why companies might want to sell their data rather than give it away.

Reliability

Labs need to provide reliable results. Their machines need to be calibrated carefully – tested on known standard samples and adjusted so that they are accurate and give repeatable results with very little variation.

Modern laboratories are very 'high-tech' and rely on automated systems. These system are:

- *fast* – using automated systems samples can be tested continuously, one after the other, 24 hours a day!

- *accurate* – once calibrated the machines will give repeatable and reliable results. Even so, they will need to be recalibrated regularly, to make that the settings have not slipped during use.

- *sensitive* – these systems can deal with microscopic amounts, or detect chemicals that make just a tiny fraction of a complex mixture.

The machines are connected to computers which do the 'number crunching', produce the charts and help with the identification. This takes away much of the boring repetition that can lead to human error, which allows the human operators to concentrate on what we can learn from the data.

Question

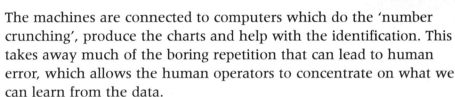

d *Environmental scientists use similar machines. What chemicals might they be looking for?*

Even with sophisticated high-tech systems errors can still creep in:

- machinery can malfunction
- human error might lead to incorrect calibration
- mistakes might be made with samples mixed or muddled up before testing
- printouts might be read incorrectly or accidentally misinterpreted
- data might be maliciously falsified for political or commercial reasons.

Because of this, the chemical data is usually cross-checked at two or more independent laboratories before any important (or expensive) decisions are made.

Key points

- Chromatography and other chemical analysis can help us find out what additives are in our food.
- Measuring instruments need to be calibrated so that they give accurate readings.
- Repeated readings are needed to give reliable measurements when averaged out.

Earth and atmosphere

Ever since people have had accurate maps of the world they have wondered at how some of the continents seem to fit together like a jigsaw.

Moving continents

People have also wondered at why the same types of land animals are found on different continents. Why are marsupials such as possums found in South America and Australia yet nowhere else?

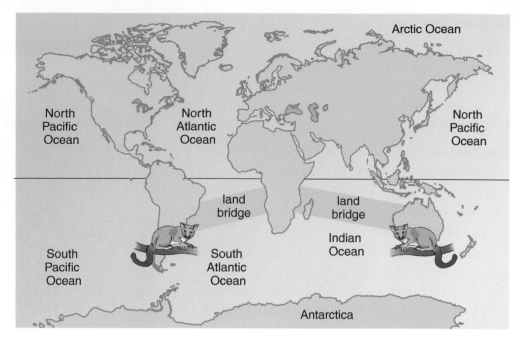

How could they have crossed deep oceans to get from one continent to another? Some scientists thought there must once have been land bridges that linked the continents. They knew that mountains rose up and then were worn away over time. Perhaps this had happened in the oceans between the continents.

In 1912, a scientist called Alfred Wegener looked closely at South America and Africa. The shapes fitted together like the two halves of a torn picture.

- Ancient mountain ranges that stopped at the coast of South America continued across the Atlantic in Africa.
- The same rocks and fossils were found on both sides.
- There was evidence of an ancient shoreline that matched.

He thought they must been part of a bigger continent that had split and moved apart. He called this process **continental drift**, but he didn't know how it worked.

When his work was published in English translation in 1924 it immediately caused uproar. Most scientists thought the idea was nonsense.

- It went against the current ideas about land bridges.
- Wegener could not explain why the continents moved.
- Errors in his data made him suggest that the continents were moving apart at 250 cm every year – that's 100 times faster than they are actually moving.
- Physicists said (rightly) that it would be impossible for continents to 'plough through the crust' without breaking up.

Today

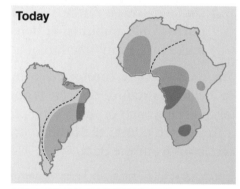

Millions of years ago

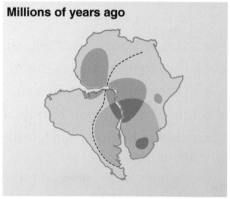

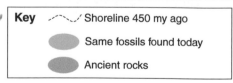

Key	⌐⌐⌐⌐ Shoreline 450 my ago
	Same fossils found today
	Ancient rocks

▲ These maps show how the continents are today and how Wegener thought the land must have been millions of years ago.

The scientific community shut Wegener out. He couldn't even get a job as a professor in Germany. He had his 'great idea' too soon, without enough evidence to back it up. He died in 1930, frozen to death on an expedition over the Greenland ice cap. He left one or two supporters, but most scientists stuck with the 'old' idea of stable continents and land bridges.

Unexpected mountains under the sea

One of Wegener's problems in getting his ideas accepted 80 years ago was that nobody knew what was at the bottom of the oceans. When scientists began a survey of the floor of the Atlantic Ocean, which is up to 7 km deep, they were amazed to discover a chain of undersea volcanoes running down the middle of the ocean! No one expected there to be mountains under the sea. In the 1960s they realised that the line of volcanoes was along a crack in the crust. Magma from the mantle pushed into this crack, cooled and set to rock. The sea floor was spreading apart and the ocean was getting wider. This was how the continents moved apart.

The seal of approval

In 1964 the Royal Society in London announced that it had enough evidence to support Wegener's idea of continental drift. Wegener's explanations were not accurate, as the whole crust moved, not just the continents. The continents were on plates made of bits of crust. But his idea finally gave birth to the theory of **plate tectonics** which revolutionised the study of the Earth.

▼ Iceland sits astride the mid-Atlantic rift. The central valley gets wider every time its volcanoes erupt.

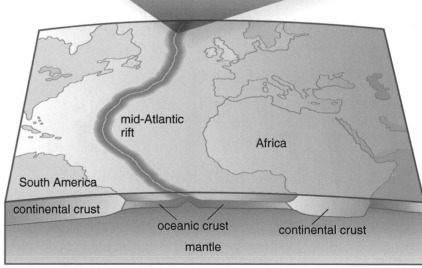

Think about what you will find out in this section

Why can it take a long time for some scientists' ideas to become widely accepted?

Why is the Earth so 'restless'?

Can we use our understanding of Earth's structure and processes to help save lives?

How has the Earth and its atmosphere changed over time?

Can we use science to reduce harmful effects of human activity on the atmosphere?

Unchanging Earth?

Living in Britain it is easy to think that the Earth is a peaceful place geologically, but don't be fooled. Over millions of years continents move around and mountains rise and are worn away. Even on a human timescale, many parts of the world are shaken by terrible earthquakes or menaced by explosive volcanoes.

Question

a Where have you heard of earthquakes occurring or volcanoes erupting recently?

Mountain building

Mountain ranges such as the Himalayas are built from folded rocks. A hundred years ago, some scientists thought that the Earth must have shrunk as it cooled down. They thought that this caused the crust to get wrinkles, just like a plum turning into a prune. They thought mountains were just big wrinkles! To understand how mountains are really made, we need to know more about the inside of the Earth.

Question

b Some mountains are huge and still rising while others are just eroded stumps. Does this fit the 'wrinkle' theory? Explain your answer.

What is the Earth like below the surface?

We live on the surface of the Earth. This surface layer is called the **crust** – it is a thin hard layer.

Beneath the crust is a hot rock layer called the **mantle**. This can move slowly at a rate of just a few centimetres a year. The continental crust 'floats' on the denser rocks of the mantle. The continents on the crust and the top part of the mantle are divided into a number of large pieces called **tectonic plates**. These move, carrying the continents with them.

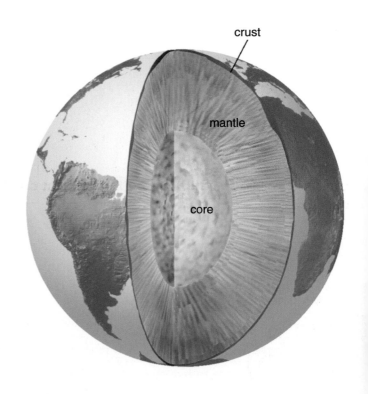

Question

c Which scientific objection to Wegener's idea of continental drift does the concept of the moving mantle overcome?

Why do they move?

If you heat a liquid from below, warm currents rise up and make the liquid swirl and mix. These are called **convection currents**. The rocks in the mantle and core are heated by natural radioactivity. This heating causes convection currents in the mantle, too. They are very slow currents but they make the tectonic plates move. The plates move very slowly, just a few centimetres a year, but they have been moving for hundreds of millions of years. Over such a long time, these tiny movements can make continents move around the Earth and force new mountains high into the air.

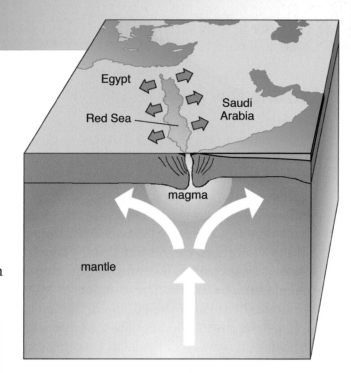

Questions

d Iceland is a volcanic island in the northern part of the Atlantic. The mid-Atlantic rift runs right through the middle. Iceland is now 2 m wider than it was 100 years ago. What is the spreading rate for the Atlantic?

e The Atlantic Ocean is approximately 5000 km wide. How old must it be if it is getting wider by 2 cm every year?

Most of the continents are now moving apart, carried along by their plates. That means that if you go back in time, they must have been closer together. Go back 250 million years and the world looked very different. All the continents were stuck together in a supercontinent called Pangaea.

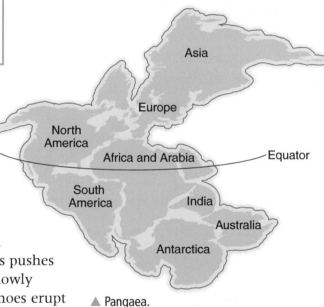

▲ Pangaea.

Mountain building cycle

In the Atlantic the plates are moving apart. In the Pacific, on the opposite side of the Earth, they are moving together. This pushes the old ocean crust back down into the mantle, where it is slowly recycled. Rocks are folded up to form new mountains, volcanoes erupt and powerful earthquakes shake the ground. Eventually old oceans will disappear completely in this way. Great mountain chains like the Alps and the Himalayas formed when oceans disappeared and the continents on each side collided. The sediment caught between the continental blocks was squeezed and folded.

▼ Mountain building where plates collide.

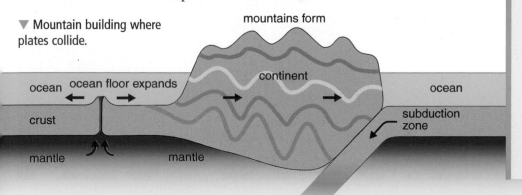

Key points

- Scientists used to think that the Earth's surface features were the result of the crust shrinking as the Earth cooled down.
- The Earth's surface consists of tectonic plates, carrying continents, which move by convection currents.
- When the plates crash into each other, they force up new mountains.

Boxing Day 2004

Every now and then the news brings stories of devastating earthquakes or erupting volcanoes. Most of these centre on uninhabited regions, and so cause few problems, but in 1999 a terrible earthquake struck Izmit in Turkey, killing 40 000 people. In 1976 an earthquake ten times as powerful killed 600 000 people in China. Volcanoes can also be deadly, though they often give more warning and so give people time to escape. The eruption of Mount Pinatubo in 1991 in the Philippines killed 350 people, but 25 000 people died after the eruption of Nevado del Ruiz in Colombia in 1985, which sent a giant mudslide down onto an unsuspecting town.

At 8 am local time on Boxing Day 2004, just off the northern tip of Sumatra in the Indian Ocean, the Earth moved. Pressure that had been building up for centuries finally caused rock to shatter and slip, triggering a massive earthquake. It was the most powerful quake for 40 years, giving out a shock wave as powerful as a thousand nuclear bombs.

The Earth's crust moved along the boundary between the Indian and Eurasian plates. One side was lifted by 10 metres – and so was the ocean above it. The water slumped back creating a huge surge wave, or **tsunami**, which spread across the ocean as fast as a jet plane.

In deep water the tsunami was not a problem. But as it approached the shore the moving water bunched up, raising the wave into a solid wall of water up to 10 metres high.

From the shore the first sign of this impending doom was a sudden drop in sea level, as if the plug had been pulled from the ocean. The unwary may have walked out to view the strange spectacle of fish stranded on the sand or suddenly exposed rock pools or reefs. And then the roar in the distance and the first sign of the tsunami wave approaching, moving faster than you could run … Within hours 250 000 people had been killed.

Question

a Earthquakes send shock wave vibrations out through the rocks. Soon after the Sumatran earthquake, elephants in Thailand trumpeted and rushed inland to safety. Suggest one way could they tell that something was wrong.

Where do quakes and eruptions occur?

Earth movements cause these sudden and disastrous earthquakes and volcanoes. Powerful earthquakes only occur along the plate boundaries. The destructive ones occur where the plates are crashing into each other like cars colliding head-on and crumpling or sliding sideways past each other.

Earthquakes and volcanoes are common. If you plot them on a map they occur in clear belts across the globe:

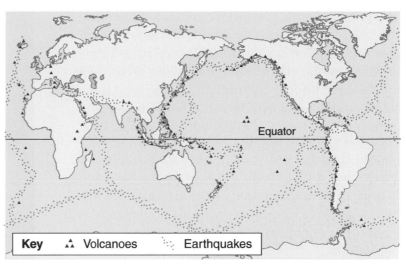

- along the 'ring of fire' around the Pacific Ocean
- along the Alps and Himalayas belt where new fold mountains are created
- in zones such as the San Andreas Fault zone in California
- in the middle of the Atlantic and Pacific Oceans where undersea volcanoes are forming
- where one plate is slipping down under another and disappearing.

▲ Earthquakes and volcanoes occur where plates meet.

Earthquake-proof buildings

Britain has no powerful earthquakes but we have the occasional weak ones. We are a long way from the edge of the Eurasian plate. Our traditional brick houses would simply shake apart in a powerful quake. Reinforced concrete should fare better, but many of the deaths in Turkey and China were caused by the collapse of poorly built buildings. San Francisco is a city of skyscrapers built over a major earthquake-prone plate boundary. Oddly enough, steel-girder framed skyscrapers survive earthquakes quite well, though they do sway alarmingly.

▼ These skyscrapers are built to withstand a good shake!

Questions

b Earthquake power is measured on the Richter scale. Each number is ten times as powerful as the number below. Izmit was magnitude 7 on this scale. What number would you give to the Chinese earthquake in 1976?

c The Sumatran earthquake was magnitude 9 on the Richter scale. Would the death toll from the Sumatran earthquake have been lower or higher if it had happened on land, under a big city?

Question

d Japan lies in the Pacific 'ring of fire'. Why do you think that Japanese houses were traditionally built with bamboo and paper walls?

Key point

- Tectonic plate movements can be sudden and disastrous, causing earthquakes and volcanoes at plate boundaries.

Vesuvius sleeps

The great volcanic cone of Vesuvius looms over the city of Naples in Italy. Nearly 2000 years ago, a gigantic eruption sent a cloud of red-hot gas down on the thriving city of Pompeii, killing all its 5000 inhabitants. Should the million people who live in Naples today be worried? You'd better believe it!

Vesuvius last erupted in 1944, when lava stopped just on the edge of the city. Since then it has been quiet – but it is far from dead. Vesuvius has a long history of dormant periods like this, often followed by catastrophic eruptions. Without doubt there will be another major eruption. The question is not 'if' but 'when'.

▲ A killer on the doorstep … if history is right.

> **Question**
>
> **a** What was different about the 1944 eruption, compared with the one that destroyed Pompeii?

When will it happen?

If the people of Naples are to have a chance of survival, they need to have advance warning so that they can evacuate the city. Scientists must try to predict when the volcano will blow, and that is far from easy. But there are some warning signs.

- *increased temperature* – There is a great chamber full of molten magma beneath a volcano, and hot gases from this escape into the crater. The temperature of this gas is about 90 °C – hot enough to cook eggs. If the temperature rises, it might mean an eruption is coming soon.
- *more earthquakes* – You get a lot of small earthquakes around a volcano. These earthquakes get stronger and more common before an eruption. Hundreds of small earthquakes have occurred recently, making it the most active period since the last eruption.

- *more gas* – The gas that comes out of a volcano is mostly carbon dioxide and toxic, foul-smelling hydrogen sulfide. You get more gas before an eruption.
- *rising land* – As magma pushes in below the volcano, the ground rises. A harbour near Vesuvius has risen nearly 4 metres out of the sea in the last 30 years.

For earthquakes, these warning signs are:

- *animal behaviour* – Local tales often talk of animals behaving strangely just before an earthquake: dogs start to bark or all the birds go silent. In the Sumatran earthquake, elephants stampeded inland to safety.
- *pre-shocks* – Minor shocks often occur before a big shock. Perhaps these are what make the animals behave strangely.
- *changes in the water levels in wells* – This often seems to drop in the period before an earthquake.

> **Questions**
>
> **b** La Sulfatara is a small crater to the West of Naples. Sulfurous gas is escaping at 165 °C. The region has had lots of small earthquakes recently and the crater floor has risen by a few metres. There is a large housing estate built on the rim. Write a letter to the local residents' association, warning them of the dangers.
>
> **c** What are the 'natural' signs that an earthquake may be on its way?

Time to evacuate?

All this activity shows an earthquake or eruption is about to occur. Will it happen tomorrow, in a few months or not for years? The best way to save lives is to evacuate the danger zone. But do this too soon, and people will get bored and drift back to their homes.

Scientists are working hard to find ways of making precise predictions about eruptions, but it is not easy. Some volcanic craters remain open, so it is relatively easy to monitor what is happening. You can see whether the lava level is rising or falling, whether it is getting hotter or cooling down, whether there is more or less gas. But with many of the more dangerous volcanoes lava sets solid in the vent, plugging it and stopping gas escaping. Building pressure eventually blows this out in a massive eruption, but it is hard to tell *how much* pressure is building up inside and *how strong* the lava 'plug' is. And there are many thousands of volcanoes in the world. Some are near large population centres in rich countries and so are monitored carefully. But poorer countries do not have the resources to do this.

For earthquakes, as you have seen there are a lot of plate edges and ocean ridges to monitor. Even if you know how they move they don't always move steadily and the stress points may be deep underground or at the surface. Since the Sumatra tsunami, scientists are considering installing an early warning system for tsunamis so people can at least be evacuated from the low-lying land.

There is a random unpredictability about many natural phenomena. Some people think it will never be possible to predict accurately the time of an eruption.

How big is the problem?

Some volcanoes have bubbling lava in their craters, which simply spills over the edge during eruptions. Vesuvius is far more dangerous as the lava has set to form a plug in the crater neck. The pressure builds up and up inside until the plug 'blows'. Look at the data table for Mount Vesuvius.

Year of eruption	Years since last eruption	Millions of cubic metres of lava erupted
1794	34	27
1858	64	120
1872	14	20
1906	34	80
1929	23	12
1944	15	25
20??		

Questions

d If you blow up a balloon and keep blowing it will eventually burst. You cannot predict the exact moment of bursting because the rubber is not perfectly uniform. If there is just a tiny flaw here, or a thinner patch there, that is where the break will begin. Use this or any other analogy to explain why it may never be possible to predict the eruption time of a volcano exactly.

e (i) Plot a graph of the amount of lava against the time between eruptions. Draw a line of best fit.
(ii) What pattern does this graph show?
(iii) Use this line to predict the volume of lava produced if Vesuvius erupted today.
(iv) Given where the 1944 lava stopped, should the people of Naples be worried? Explain your answer.

Key point

- Although scientists monitor for earthquake and volcano activity, they find it very hard to predict accurately when they will occur.

What's in the air?

The Earth's atmosphere has been more or less the same for 200 million years. Without it, we could not exist. Dry air is 21% oxygen, the gas you need to breathe; 78% is the unreactive gas nitrogen. There is just 1% of 'other gases'. Normal air also contains varying amounts of water vapour. The 21% of oxygen is very important to animals on the Earth. The carbon dioxide is very important to plants.

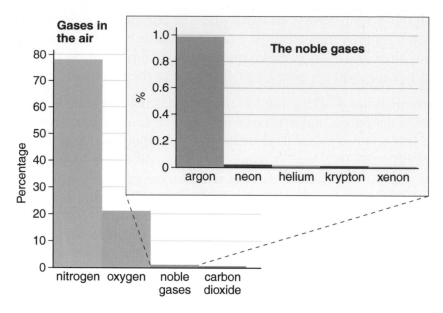

What are the 'other gases'?

A hundred years ago, nobody knew that these 'other gases' existed. They were hidden in the fairly unreactive nitrogen. But nitrogen reacts with burning magnesium. Careful experiments showed that there was 1% of air that refused to react even with magnesium. This was called **noble gas**, as it was so unreactive. (Unreactive gold is called a noble metal.) We now know that there is a whole family of noble gases. The most common is argon.

The noble gases form Group 0 of the periodic table. The noble gases are typical non-metals in many ways:

- they have very low melting/boiling points
- when solid they are soft and crumbly
- they do not conduct electricity.

They are, however, very unreactive. You may think that such unreactive gases will not be of much use, but sometimes their very inactivity is just what is required.

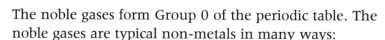

Group 0

He
Ne
Ar
Kr

Question

a Oxygen boils at −183 °C, argon at −186 °C and nitrogen at −196 °C. Suggest another way that argon could be separated from air.

Using the noble gases

Helium is not very soluble in water, even under pressure. This makes it an excellent substitute for nitrogen in the 'air' that deep-sea divers breathe. When a diver surfaces nitrogen bubbling out of the blood can cause the 'bends'. Helium doesn't do this but does have the side effect of making you sound like Donald Duck when you talk!

Question

b Methane (in natural gas) boils at −162 °C, while helium boils at −269 °C. Explain what would happen if you cooled natural gas to −180 °C. Why do you think it is easier to separate helium completely from natural gas than it is to separate argon from air?

Helium's low density means that it is very easily lost from the atmosphere, into space. Helium forms in the rocks of the Earth during radioactive decay. For the last 50 years or so, helium has been produced commercially from natural gas.

Helium's low density also makes it a safe alternative to hydrogen for modern airships – or party balloons.

Question

c (i) Why is helium a safer gas to use in airships than hydrogen?
(ii) Why wasn't it used for the early airships of the 1930s?

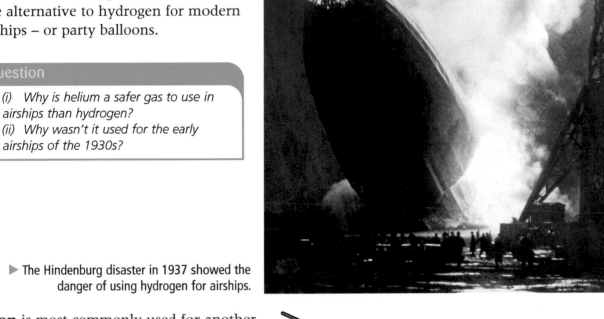

▶ The Hindenburg disaster in 1937 showed the danger of using hydrogen for airships.

Neon is most commonly used for another property. It glows if an electrical discharge is passed through it. These discharge tubes can be coloured, making them ideal for flashy neon signs.

Argon is the cheapest of the noble gases to produce as it makes up 1% of the air. Argon is used inside ordinary light bulbs, to stop the filament from burning. It is also used to give an unreactive atmosphere for welding which could be dangerous if done in air.

Question

d Why is argon used rather than helium or neon?

Krypton is used in some lasers. These are used for laser surgery and for removing birthmarks and tattoos, as well as for 'laser sights' on rifles. They can also be used to cure some eye defects.

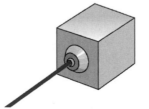

Key points

- For 200 million years the Earth's atmosphere has consisted of roughly 80% nitrogen and 20% oxygen. There is a small proportion of carbon dioxide and water vapour and noble gases.
- Noble gases are in Group 0 of the periodic table and are chemically unreactive, which can be a useful property.

The first atmosphere

The Earth formed 4½ billion years ago. In the beginning things were very different from our present atmosphere of oxygen and nitrogen. The Earth was very hot and was covered with volcanoes. Gas from these volcanoes formed the first atmosphere. This first atmosphere was thought to be made mostly of carbon dioxide and water vapour, with a little methane and ammonia. Scientists have come up with the theory of the origin of the atmosphere by studying the gases from modern volcanoes, as well as by looking at the atmospheres of Mars and Venus.

▲ The Earth 4.5 billion years ago may have looked like this.

Questions

a Near Naples in Italy there is a cave known as 'the mouth of hell' that fills up with invisible, odourless volcanic gas. Any dogs that wander in die by suffocation. What gas causes this, do you think?

b Which type of organism wouldn't have been able to survive in a carbon dioxide atmosphere?

As the Earth cooled the water vapour turned to water, forming the oceans. The atmosphere was almost 100% carbon dioxide, just like on Mars and Venus today. We know that there was no oxygen as iron found in the sediments is not oxidised.

Simple microbes lived in this oxygen-free environment. Then about 3 billion years ago simple plants evolved in the oceans. These plants changed the world. Fossils show that they became bigger and more sophisticated. Eventually they colonised the land as well.

▲ 3 billion years ago.

The first pollutant on Earth

Plants use photosynthesis to make the food they need from carbon dioxide and water. They do this by tapping into the vast amounts of energy that pour onto the Earth from the sun. This gave them a competitive edge over the pre-existing organisms, so plants grew, evolved and spread rapidly through the oceans. But plants make oxygen when they photosynthesise. To the simple microbes, oxygen was a poison! So the growth and spread of plants led to a pollution of the world's oceans with oxygen because oxygen dissolves in water. You can see evidence for this in the thick beds of oxidised iron that formed at this time.

Question

c Imagine a new form of life evolved that produced chlorine gas. What effect would that have on the microbes, plants and animals that live on Earth today?

▲ 2 billion years ago.

By 2 billion years ago, early life forms were nearly wiped out. Just a few survived in oxygen-free deep ocean mud or stagnant pools.

Locking up the carbon

As plants evolved and spread, carbon dioxide was removed from the atmosphere and locked up in the plant biomass. Some of this became trapped in the rocks, forming the fossil fuels of the future. By 300 million years ago great swamp forests covered much of what is now Britain. You can find the stumps of fossil trees in many parts of the country.

Meanwhile, oxygen started to build up, first in the oceans and then finally in the atmosphere. Just under 1 billion years ago the first animals appeared in the oceans, exploiting the 'new' oxygen to get the energy they need for life by respiration.

Some of these ocean organisms built shells from calcium carbonate. They took carbon dioxide from the oceans for this, so more dissolved out of the air to take its place. When they died their shells formed limestone which locked the carbon away in the rocks. So carbon dioxide levels in the atmosphere fell as oxygen levels rose. By about 200 million years ago, the oxygen and carbon dioxide levels reached their present values and the atmosphere has remained fairly stable ever since.

The graph shows how the percentage of carbon dioxide and oxygen in the atmosphere has changed over time.

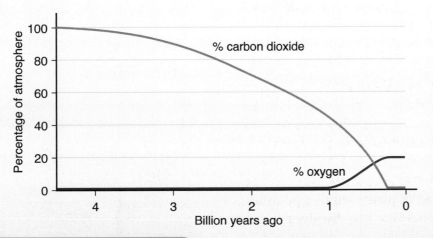

Question

d Why couldn't animals have evolved before plants?

Questions

e Every tonne of coal will have removed 3.7 tonnes of carbon dioxide from the atmosphere. What happens if that coal is burnt?

f Every tonne of limestone will have removed 0.44 tonnes of carbon dioxide from the atmosphere. Limestone is a very much more common rock than coal (coal beds are just a few metres thick, limestone beds may be hundreds or even thousands of metres thick). Which has removed more carbon dioxide from the atmosphere, coal or limestone? Explain your answer.

Question

g (i) For roughly what proportion of Earth's history has the oxygen level been as it is today?

(ii) Simple plants had been producing oxygen for 2 billion years before it started to build up in the atmosphere. Where did the oxygen go over that first period?

Key points

- In the first billion years of the Earth's existence the atmosphere was very different from today.
- Theories suggest that the early atmosphere was mainly carbon dioxide, until plants developed and produced oxygen.
- Most of the carbon dioxide from the early atmosphere is now locked up in fossil fuels and limestone rock.

Tipping the balance

For 200 million years our atmosphere has been in balance. Plants take in carbon dioxide and animals breathe it out. Recycling at its best! But over the last 200 years we have upset the balance. We've been burning up the fossil fuels a million times faster than they took to form. What effect will this have on the Earth?

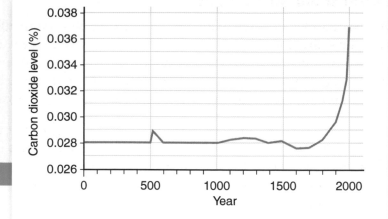

▲ Carbon dioxide level in the atmosphere over the last 2000 years.

Question

a Look at the graph.
(i) Describe in words what the graph tells you.
(ii) There was a huge volcanic eruption in 500 AD. What effect did this have on the carbon dioxide level?
(iii) How long did this effect last – 50, 200 or 500 years?
(iv) How long has the recent rise in carbon dioxide level been going on for?
(v) Do you think this could have been caused by volcanic activity? Explain your answer.
(vi) What human activity over the last 200 years might have caused this?

Theories and evidence

Environmentalists are certain that the Earth's climate is changing and point the finger of blame at fossil fuels. Most scientists agree that the climate is changing and that the billions of tonnes of carbon dioxide that we put into the air every year when we burn fossil fuels is contributing to this change, but this may not be the sole cause. There are several different theories about what effect human activity is having on the Earth's atmosphere now. Beyond that, there is little agreement.

Carbon dioxide to blame?

Environmentalists believe we need to reduce the amount of carbon dioxide produced to try to stop the climate from changing. In a meeting in Kyoto in 1997, leaders of the developed world, including Britain, agreed to set targets for reducing the amount of carbon dioxide they produce.

But the biggest polluter of all, the USA, has not agreed to this. They say using less oil could harm their economy and people would suffer. They also point out that India and China, both countries with huge populations, are industrialising fast and will soon produce more carbon dioxide than even the USA.

Question

b Look at the table below.
(i) From these figures, calculate the total mass of carbon dioxide produced every year by each country.
(ii) Over the last 20 years, the figure for the USA has been constant, while the figures for India and China have doubled. Suggest a reason for this.
(iii) What would be the overall effect on global carbon dioxide emissions if this trend continued for another 20 years?
(iv) Would a 15% reduction by the USA over this period be significant for the environment?

Country	Population	Tonnes of carbon dioxide produced per person annually
USA	300 million	20
China	1300 million	3
India	1000 million	1

Natural climate change

Environmentalists often give the impression that the climate would stay the same if it were not for us. Four hundred years ago, Britain was so cold in winter that the River Thames froze over in London and fairs were held on it. Eight hundred years ago it was much warmer than today, and grapes were grown to make wine in northern Britain.

Going further back, 20 000 years ago the world was in the grip of the Ice Age. Britain was covered by great ice sheets and woolly mammoths roamed the country. So much water was locked up in the ice that the sea level fell so far that you could walk to what is now France. Then 10 000 years ago the ice started to melt and sea levels rose, flooding communities that lived in low-lying areas.

Research shows that the Earth's climate has been alternating between ice age and much warmer periods when all the ice melted every 100 000 years or so for millions of years. No-one is quite sure why, though cycles of solar activity or passing interstellar dust clouds have been blamed. Our climate is currently about halfway between the two extremes.

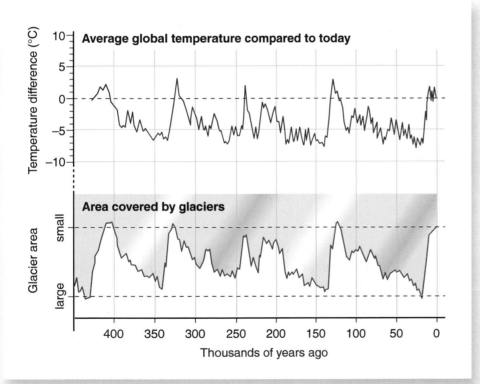

Key points

- We are releasing carbon dioxide locked up in fossil fuels from the Earth's early atmosphere and this is increasing the carbon dioxide level in today's atmosphere.
- We can evaluate theories about the changes occurring in today's atmosphere by looking at the evidence.
- We can evaluate the effects of human activity on today's climate by analysing data.

Questions

c Why is it extremely unlikely that the Earth's climate will stay constant?

d From the graph, which is usually more rapid, the onset of an ice age or its retreat?

1 a Which of the following shows a saturated hydrocarbon?

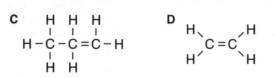

b Which of the following could represent an alkene?

A C_5H_{12} B $C_{16}H_{34}$
C $C_{18}H_{36}$ D $C_{42}H_{86}$

c Unwanted long-chain hydrocarbons are
  A cracked to make the monomers needed to make addition polymers
  B combined with bromine water to make polymers
  C are used as the monomers to make addition polymers
  D converted to alkanes to make addition polymers

d Small alkene molecules are
  A cracked to make the monomers needed to make addition polymers
  B combined with bromine water to make polymers
  C are used as the monomers to make addition polymers
  D converted to alkanes to make addition polymers

e Some people think plastic polymers are an environmental hazard because they
  A are biodegradeable
  B are not biodegradeable
  C are easily moulded into shape
  D are not easily moulded into shape.

2

Polymer	Properties
poly(propene)	semi-rigid, easily coloured
poly(ethene)	cheap, flexible
poly(styrene)	rigid, easily moulded
PVC	tough, resists damage, shiny, electrical insulator
PET	transparent or black, microwave-proof

For each use below, suggest a suitable plastic and give a reason why.

a a supermarket carrier bag   *(1 mark)*
b the packaging for a CD player   *(1 mark)*
c 'artificial leather' for a sofa   *(1 mark)*
d a washing-up bowl   *(1 mark)*
e a lemonade bottle   *(1 mark)*
f coating for an electric cable   *(1 mark)*
g the tray for a microwave dinner   *(1 mark)*

3 The graph shows how much energy you get by burning 1 kg of plastic waste compared to 1 kg of oil and coal.

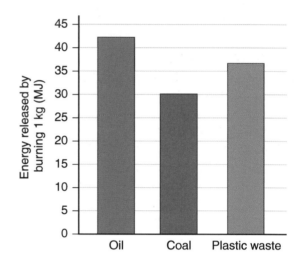

a Describe how the energy from plastics compares to the energy from fossil fuels.
    (1 mark)

b i Most waste plastic in Britain ends up in landfill sites. Give **two** reasons why this is not a good way to deal with plastic waste. *(2 marks)*
  ii Suggest a reason why it might be difficult to recycle plastic waste. *(1 mark)*

c i We are responsible for up to 0.5 kg of waste plastic every day. Suggest **two** advantages of burning this material in power stations instead of coal or oil. *(2 marks)*
  ii What are the **two** main compounds that will be formed when plastic burns?
    (2 marks)

d Some plastics contain nitrogen or chlorine. Why might these cause a problem? *(1 mark)*

4 A third of all plastic used goes in packaging – which is then simply thrown away. Manufacturers now stamp many plastic containers with the plastic type to aid recycling.

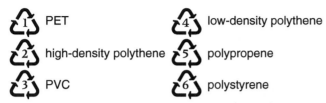

1 PET 4 low-density polythene

2 high-density polythene 5 polypropene

3 PVC 6 polystyrene

a i How easily would it be to separate out mixed plastic waste using these codes (could it be automated)? *(1 mark)*

ii Some councils have recycling bins to allow people to sort their waste by hand into the different groups. Suggest a way to encourage people to do this. *(1 mark)*

Some supermarkets are experimenting with carrier bags made using polymers made from corn starch that are biodegradeable.

b i Suggest **two** ways in which these would be better than polythene. *(2 marks)*

ii Suggest **two** possible problems that might stop the general introduction of these bags. *(2 marks)*

Scientists are now looking at ways to break down waste plastics into smaller chemicals which could then be fractionated, cracked and used as 'feedstock' for new plastics. The problem is that expensive equipment is needed and large amounts of plastic waste would need to be available to make it economically viable (at least 50 000 tonnes a year).

c i Why is it easier to separate out oils and other chemicals formed by breaking down plastics than it is to separate the plastics themselves? *(1 mark)*

ii Suggest **two** factors that are slowing down the introduction of this technology. *(2 marks)*

5 Match words **A**, **B**, **C** and **D** to the sentences in the table.

    A whipped cream   **B** egg yolk
    C cream   **D** butter

1	This is a 'water' in oil emulsion made by churning milk.
2	This oil in water emulsion separates out from milk.
3	This oil in water emulsion has added air bubbles.
4	This is added to oil and vinegar in mayonnaise as an emulsifier.

6 Read this passage and answer the questions.

> Oils are a much more concentrated energy store than carbohydrates. Many plants store oil in their seeds, to provide the energy for the growing seedling. When we eat vegetable oils, we can utilise this energy store to give us the energy we need for life. There is a concern that eating too much saturated oil could lead to the build up of cholesterol in the arteries, which can lead to heart disease. But many vegetable oils are unsaturated, and these seem to have the opposite effect, keeping us healthy.
>
> We all need to have some oil in our diet, as it is needed for important reactions in our cells. Fortunately we can make most of the oil we need, but some oils – called omega-3 and omega-6 – have to come from our diet. Omega-3 oils are particularly good for our brains. They are found in fish oils, but also in some vegetable oils such as flaxseed oil.

a i Explain the terms saturated and unsaturated. *(2 marks)*

ii How could you use bromine water to tell these two types of oil apart? *(1 mark)*

b i Which chemical causes heart disease if it builds up in the arteries? *(1 mark)*

ii Which types of oil seems to prevent this effect? *(1 mark)*

iii One 'old wives' tale' was that 'fish is good for your brain'. Could there be any truth in this? *(1 mark)*

iv How can vegetarians keep their brain in good shape? *(1 mark)*

7 Vegetable oils are usually unsaturated, but animal fats are saturated.

  a What health problem is associated with eating too much animal fat? *(1 mark)*

  b Olive oil is a healthy oil for salads but not so good for frying as heating for a long time breaks down the C=C double bonds in unsaturated oils. In what way does this 'spoil' the oil as a food? *(1 mark)*

  c **i** Vegetable oils are turned into solid fats for margarine by bubbling hydrogen through the hot oil. What happens in this reaction? *(1 mark)*

    ii Nickel is added to this mixture. What effect does it have on the reaction? *(1 mark)*

8

Fuel	Energy content (MJ/litre)	Source	Main pollutants produced compared to petrol engine *		
			Particulates	Nitrogen oxide	Carbon monoxide
petrol	26	crude oil	*	*	*
diesel	29	crude oil	high	*	*
biodiesel	27	vegetable oil	low	high	*
ethanol	18	sweetcorn	*	low	low

  a **i** How does biodiesel compare to petrol and diesel in terms of energy content per litre. *(1 mark)*

    ii How does ethanol compare to petrol and diesel in terms of energy content per litre. *(1 mark)*

  b Cars can be converted to run on either petrol or ethanol. Which do you think would give more miles per litre? *(1 mark)*

  c **i** City smog is caused by a mixture of nitrogen oxides, carbon monoxide and particlulates. Is ethanol more or less likely to cause city smog. *(1 mark)*

    ii What pollution disadvantage does biodiesel have compared to ethanol. *(1 mark)*

    iii What big advantage do biodiesel and ethanol have over petrol and diesel in the long term? *(1 mark)*

  d You would need a 4000 m² field to produce enough biodiesel to run a car for a year. There are 24 million cars in Britain. What is the big limitation on using biodiesel to replace fossil fuels in Britain? *(1 mark)*

9 The ideas about how mountains form have changed dramatically over the last 100 years.

  a How did early scientists think that mountains were formed as the Earth cooled? *(1 mark)*

  b We now know that the crust is broken up into slabs called plates that move slowly. What makes these plates move? *(1 mark)*

  c Describe how moving plates can cause mountains to form. *(1 mark)*

  d The edges of plates often stick and jam, but eventually they will jerk apart and move. What does this cause? *(1 mark)*

10 Yellowstone Park is a vast area in the Rocky mountains of Wyoming, USA. It is famous for its geysers that squirt boiling water into the air in great fountains. It also has bubbling pools of mud and sulfurous fumaroles. Recently there have been many small earthquakes, and the ground has risen by a metre or so in some areas.

  a Give **two** features that suggest that Yellowstone is a volcanic region. *(2 marks)*

  b Give **two** features that suggest an eruption might be on the way. *(2 marks)*

Scientists think that Yellowstone Park contains a 'supervolcano' that, when it blows, would be a thousand times as large and powerful as any other eruption in recorded human history. It would produce 1000 times as much ash, carbon dioxide and sulfur dioxide.

  c Suggest some effects that this type or eruption might have on the Earth's climate. *(2 marks)*

11 This question is about the atmosphere and how it has changed over Earth's history.

  a The composition of the air today is approximately

    A 1/5 nitrogen and 4/5 oxygen

    B 1/5 carbon dioxide and 4/5 oxygen

    C 1/5 oxygen and 4/5 carbon dioxide

    D 1/5 oxygen and 4/5 nitrogen.

b Four billion years ago the Earth's atmosphere was mostly made from

   A nitrogen

   B oxygen

   C methane

   D carbon dioxide.

c There is oxygen in the atmosphere because

   A animals need oxygen to breathe

   B animals make oxygen when they respire

   C methane reacts with carbon dioxide

   D plants make oxygen during photosynthesis

d There is less carbon dioxide in the atmosphere on Earth than there once was because

   A we burn fossil fuels on Earth

   B lots of carbon is locked up in limestone and fossil fuels

   C animals breathe out carbon dioxide

   D it has turned into nitrogen

e Just under 1% of the air is a gas called argon. This gas was not discovered for a very long time because

   A it is invisible

   B it just has single atoms

   C it is an unreactive gas from Group 0 of the periodic table

   D it is very similar to oxygen so it was very hard to tell the two gases apart

12

Location	Type of building
England (traditional)	brick and mortar
Japan (traditional)	wood and paper
USA (California – modern)	concrete and steel

a **i** Which of the regions shown above suffer from powerful earthquakes? *(2 marks)*

   ii What is the reason for this? *(1 mark)*

b **i** During an earthquake the ground shakes violently. What would happen to a traditional English house in a powerful earthquake? *(1 mark)*

   ii Why would this be very dangerous to people in the house? *(1 mark)*

c **i** What would happen to a traditional Japanese house in a powerful earthquake? *(1 mark)*

   ii Why would this be less dangerous to people in the house compared to an English style house? *(1 mark)*

d **i** Concrete and steel-frames buildings are flexible and can sway without breaking. Are skyscrapers more or less vulnerable than 'bricks and mortar'? *(1 mark)*

   ii Which material used extensively in modern skyscrapers might be more vulnerable to earthquake damage? *(1 mark)*

13 Water levels in the rocks appear to change before an earthquake. Look at this table from a Chinese mine, from the three weeks before the 1976 earthquake. It shows how fast water had to be pumped out to stop the mine from flooding.

Time before the earthquake (days)	Pumping rate (m^3 per second)
21	70
14	65
10	55
7	45
5	35
4	30
3	25
2	25
1	50
0	75

a Draw a graph of these figures and describe the pattern. *(4 marks)*

b Describe what would have happened to the water level in local wells over the period. *(1 mark)*

c What happened just before the earthquake? *(1 mark)*

d Suggest why water levels should be studied carefully in earthquake zones. *(3 marks)*

Atoms build matter

Nowadays, we all know that matter is made up of small particles called atoms. The idea is an old one, but it didn't catch on straight away. The Greek philosopher Democritus first suggested the idea 2 400 years ago. But his rival Aristotle, even more famous for his observations and thoughts about the natural world, disagreed. Neither tried to test their results. As a result, the idea of atoms was ignored for 2 000 years. Science can only develop if ideas are tested by experiments.

Chemistry born from alchemy

Later on, people became more interested in the practical uses of materials. A thousand years ago alchemists tried to make gold (rare and expensive) from 'base metal' such as lead (cheap and available). They failed. They had no real idea what was happening when they made chemicals react. They were following Aristotle. They just tried different combinations in the hope that they would make a great discovery. It was a form of primitive 'cookery'. They developed some interesting techniques such as distillation. And they spotted one or two useful patterns that helped later chemists. (They also blew themselves up quite often!)

John Dalton, who lived 200 years ago, figured out how to test whether Democritus was right or not. His results showed that materials were definitely made up of atoms, and that they combined in various ways to make the substances that form and surround us. He did some very careful experiments, weighing the reactants and products. He concluded that:

- every element is made of its own, distinctive atoms of a particular mass
- other chemicals are made from atoms that have joined together in some way.

▲ Aristotle – a great thinker (but not such a good scientist).

▲ An alchemist – a great (but flawed) experimenter.

He gave his elements special symbols and drew pictures of some simple compounds. He may not have got everything quite right, but he certainly got chemistry moving in the right direction.

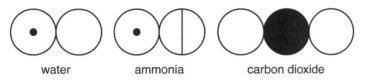

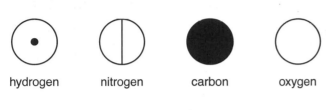

▲ Some of Dalton's elements and compounds.

Chemists take control

Today, we have a very good understanding of what is going on in a chemical reaction. With this understanding comes control. Chemists today can work out how best to get metals from their ores, or turn crude oil into thousands of useful products, for example. They can also create new superdrugs to combat disease or design fantastic new materials to keep our technology marching forward. The only limit is our imagination.

▲ John Dalton – a great experimenter, a great chemist.

▶ Modern chemists have a lot to thank Dalton for.

Think about what you will find out in this section

What is inside the atom?	How do atoms join together?
How can matter take on so many different forms?	Why are materials so different?
Why do elements have different properties?	

Naming the pieces

Everything is made up of particles. For most substances, these particles are built from smaller particles called **atoms**. Groups of two or more atoms joined together are called **molecules**.

There are about 100 different kinds of atoms on Earth. Substances made from one type of atom only are called **elements**. As there are 100 different types of atom, there are 100 different elements. Some elements, such as oxygen or sulfur, form molecules containing just one type of atom.

Compounds

When different elements combine they form **compounds**. The 100 different atoms can combine to give millions of different compounds, but the same compounds are always formed in the same way.

Inside the atom

To understand how atoms join together to make all these different chemicals, you need to know how the atoms themselves are made. At the very centre of each atom we find the tiny **nucleus**, where most of the mass is concentrated in a small volume. Small, light **electrons** whizz around the nucleus in the rest of the space.

When we look a bit harder at the nucleus, we find it is made up of two types of sub-atomic particle: **protons**, which each carry a unit of positive electrical charge, and the **neutrons**, which are electrically neutral. The charge on the protons is balanced out by the charge on the electrons, which each carry a unit of negative charge. So overall, the atom is electrically neutral.

When you look at a TV picture, there is just a moving dot of light, but it moves so fast that you see a complete picture. You are watching electrons strike the screen at a fantastic rate. Similarly, in the atom, the electrons are in constant motion around the nucleus, going so quickly it seems they are everywhere at once. The effect is to make an outer shell for the atom, which gives it a shape and makes the atom seem much larger than the nucleus. That's why we say, surprising as it may seem, that most of every atom is just empty space.

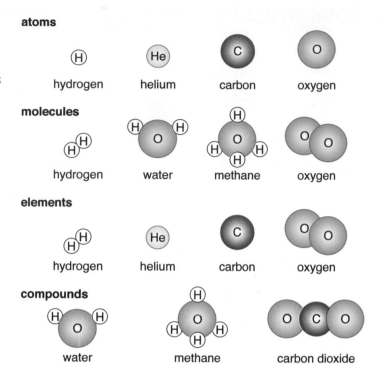

▲ Simple models for some atoms, molecules, elements and compounds.

Questions

a How many atoms of oxygen are there in an oxygen molecule?

b Which elements combine together to make water molecules and which make methane?

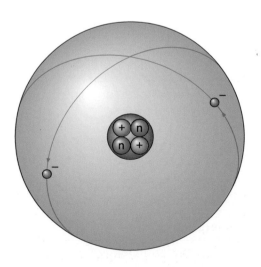

▲ The building blocks of the atom.

Sub-atomic particle	Charge	Position
proton	+1 (positive)	in the nucleus
neutron	0 (neutral)	in the nucleus
electron	−1 (negative)	whizzing around the nucleus

Which element? Count the protons

What distinguishes one element from another? Atoms of the same element always have the same number of protons in the nucleus. If the number of protons is different, it is a different element. The number of protons an element has is called its **proton number** or **atomic number** (Z). Hydrogen is the simplest atom, with just 1 proton, so it is atomic number 1. Helium has 2 protons, so it is atomic number 2. Uranium has 92 protons, so it is atomic number 92.

The electron cloud takes shape

The electrons whizz round the nucleus of an atom forming an 'electron cloud'. This gives the atom its shape. But the electron cloud is not haphazard. The electrons can only fit into certain zones. These are called **energy levels** or **electron shells**. You can use both terms, but from now on we will use mainly the first.

The electrons normally fit into levels closest to the nucleus. For the first 20 elements, the pattern is simple:

- **first level:** just 2 electrons and this one is full
- **second level:** this can take up to 8 electrons
- **third level:** this can also take up to 8 electrons
- **fourth level:** the pattern gets more complicated after the first 2 electrons in this level.

Question

g How would (i) 7 (ii) 11 (iii) 17 electrons fit into this pattern?

Questions

c Which sub-atomic particles give atoms their shape?
d In 1911, Ernest Rutherford fired tiny radioactive particles at a thin sheet of gold and found that most passed straight through. How could this happen?

Questions

e How many electrons are there whizzing round (i) a hydrogen atom (ii) a uranium atom?
f X, Y and Z are three atoms. X has 12 particles in its nucleus (6 protons and 6 neutrons), Y has 14 (6 protons and 8 neutrons) and Z has 14 (7 protons and 7 neutrons). Which two are the same element?

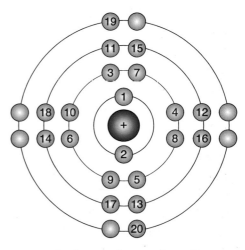

▲ How the first 20 electrons fit in the energy levels. You can mark a cross (x) to show the position of an electron.

Key points

- All matter is made up of atoms.
- Atoms have positive protons and neutral neutrons in the nucleus, with tiny negative electrons in energy levels whizzing around them.
- The number of protons is called the atomic number (or proton number). Different elements have different numbers of protons.
- Atoms are neutral as they have the same number of protons and electrons.

Filling up with electrons

Remember that atoms are neutral so they have the same number of electrons as their proton number. Here is how some atoms fill up the electron levels.

- Hydrogen's single electron fits in the first level, but helium's 2 electrons fill it up completely.
- At atomic number 3, lithium's third electron must start to fill the second level.
- Carbon has atomic number 6, so it fills the first level and half-fills the second level.
- Neon has 10 electrons which fill up the first and second levels.
- Sodium starts a third level for its 11th electron.
- Element numbers 12 to 18 fill up the third level.
- Calcium at number 20 is the last element to show this simple pattern. Its last 2 electrons are found in the fourth level.

$^{1}_{1}$H 1 $^{4}_{2}$He 2

$^{7}_{3}$Li 2,1 $^{12}_{6}$C 2,4 $^{20}_{10}$Ne 2,8

$^{23}_{11}$Na 2,8,1 $^{40}_{20}$Ca 2,8,8,2

▲ Electrons usually fit into the lowest available energy level.

Questions

a How many electrons does an atom of magnesium (Z = 12) have? How many electrons will fit in each energy level?

b Draw the electronic structure of (i) fluorine (Z = 9) (ii) aluminium (Z = 13) (iii) sulfur (Z = 16) (iv) argon (Z = 18) (v) potassium (Z = 19).

c How many full energy levels does potassium (K) have?

Making sense of the periodic table

When he drew up the periodic table in 1869, the Russian scientist Dimitri Mendeleev didn't know about electrons. He did know the pattern of chemical properties was repeated, but he couldn't explain why. See if you can do better. Compare the electronic structure of the first 20 elements to the periodic table.

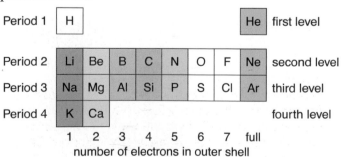

Period 1	H							He	first level
Period 2	Li	Be	B	C	N	O	F	Ne	second level
Period 3	Na	Mg	Al	Si	P	S	Cl	Ar	third level
Period 4	K	Ca							fourth level
	1	2	3	4	5	6	7	full	

number of electrons in outer shell

▲ The periodic table, with electron levels shown on the right.

d Copy the simple periodic table for the first 20 elements as shown. Label the groups and periods. For each element, mark in the number of electrons in the outer electron level, and the number of levels being used. What pattern does this show? How do the groups and periods relate to the electronic structure?

e How many electrons are there in the outer electron level of
(i) rubidium (Group 1)   (ii) lead (Group 6)
(iii) iodine (Group 6)   (iv) xenon (Group 0)?

Electrons rule chemistry

As you have shown for the first 20 elements, the pattern of the periodic table matches the pattern in the way electrons fill the energy levels. Things get complicated after calcium, as the transition metals wedge in, but the pattern does still hold. For all the elements in the eight groups of the periodic table, the period tells you the number of energy levels and the group number tells you the number of electrons in the outer level.

When atoms collide it is the outer levels that come into contact. It is perhaps not surprising then that atoms with the same number of electrons in the outer shell react in similar ways. Different elements with the same outer electron configuration therefore show similar properties, which explains why Mendeleev grouped them together.

▲ You now know more about the periodic table than Mendeleev did!

Making compounds

Most substances are compounds. The way that compounds form depends on the electrons in the outer energy levels of the atoms involved.

- A compound made of metals and non-metals is usually made by a process of 'give and take' (transfer) of electrons between outer shells of atoms.
- In a compound made of non-metals, the electrons have to be shared rather than transferred.

Question

f Sodium chloride (NaCl) is formed from sodium and chlorine atoms. From your answer to **e**, which atom do you think will 'give' and which do you think will 'take' an electron?

Key points

- Electrons fit into energy levels that fill from the 'bottom' upwards.
- The way an atom reacts depends on the number of electrons in the outer energy level.
- Elements in the same group in the periodic table have the same number of electrons in the highest energy level.

Happiness is a full energy level

Helium (Z = 2), neon (Z = 10) and argon (Z = 18) belong to a family of elements called the **noble gases**. They have this name because they keep to themselves and do not join in chemical reactions at all. What is it that makes them so stable and unreactive?

If you look at their electronic structure, you will see that the noble gases all have full outer energy levels. This appears to be a very stable arrangement, which is not easy to upset. Or to put it in a less scientific way, as far as atoms are concerned, 'happiness is a full outer energy level'. As we shall see, atoms of other elements are always trying to achieve a similar result. In the process, they often have to team up.

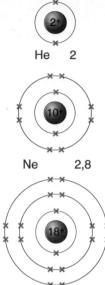

◀ The full outer energy levels of the noble gases make them very unreactive.

He 2

Ne 2,8

Ar 2,8,8

Question

a Atoms of the element krypton have 8 electrons in their outer energy levels. What predictions can you make about the properties of krypton? Explain your answer.

Fast and loose: the single outer electron

Group 1 of the periodic table contains the alkali metals, including sodium.

Question

b Name the other alkali metals.

These elements have just one electron in their outer energy level. This electron is quite easily lost. If this happens, the atom is no longer neutral. It has one more positive proton in the nucleus than there are negative electrons. This means that overall the particle now has a single positive charge. Charged particles like this are called **ions**.

A sodium ion is smaller than a sodium atom as it has one less occupied energy level. Ions like this have the electronic structure of a noble gas, so they are very stable.

◀ Ions have the electronic structure of a noble gas, but are electrically charged.

sodium atom

Question

c Draw similar diagrams to show how potassium (K) becomes a K⁺ ion.

On the scrounge: the nearly-filled electron shell

Group 7 of the periodic table contains non-metallic elements called the **halogens**. Group 7 elements such as chlorine all have 7 electrons in their outer energy level. These are tightly held, but Group 7 atoms can capture an extra electron to fill the energy level. This forms a stable ion which has the electronic structure of a noble gas. As Group 7 atoms gain an extra electron, their ions have a single negative charge. They are called **halide** ions.

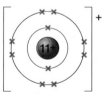

sodium ion

neon atom

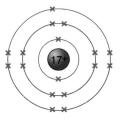

chlorine atom

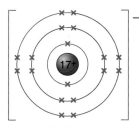

chlorine ion

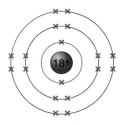

argon atom

d Aluminium is in Group 3. What is the charge on an aluminium ion?

e Atoms of iodine have 7 electrons in their outer energy levels. Is iodine a metal or non-metal and what ion does it form?

Metals and halogens: made for each other

All metal atoms have 1 or 2 (or more) electrons that are easily lost in their outer energy levels. When they lose these electrons they form positive ions. Group 2 metals form 2⁺ ions, Group 3 metals form 3⁺ ions, and so on.

Non-metal atoms have 5, 6 or 7 electrons in their outer energy levels. These are not easily lost. Instead, the atoms gain extra electrons to get a full shell of 8 electrons. Group 5 non-metals have to gain 3 electrons to form 3⁻ ions, Group 6 non-metals have to gain 2 electrons to form 2⁻ ions and Group 7 non-metals have to gain 1 electron to form 1⁻ ions. Forming such ions enables the atoms to attain the stable electronic structure we see in the atoms of the noble gases (Group 0).

Getting hitched: the ionic bond

You may have spotted the obvious connection. Metallic atoms become stable ions by losing electrons. Non-metallic elements become stable ions by gaining electrons in their highest energy levels. Together, the atoms form a compound, an entirely new substance, that does not necessarily resemble either of the elements that have come together to form it. Often, the compound is a **crystal**, a solid that consists of an enormous **lattice** of ions held together by the forces of attraction between them. This method of joining is called an **ionic bond**. The compound is called an **ionic compound**.

Questions

f Suggest a reason why metals are able to conduct electricity.

g Calcium is in Group 2. How many electrons can a calcium atom lose when it forms a compound?

Question

h What kind of bond would you get between calcium and oxygen? Why?

Key points

- Having a full outer energy level makes an atom very stable.
- Atoms with just a few outer electrons try to lose them.
- Atoms with 'nearly full' outer energy levels try to gain electrons.
- Alkali metals (Group 1) form positive ions by losing their single outer electron, leaving the energy level beneath full.
- Halogens (Group 7) form negative ions by gaining an extra electron to fill their outer energy level.

The salt on your table

Sodium is a soft and dangerously reactive metal. Chlorine is a green poisonous gas. Burn sodium in chlorine and the violent reaction leaves white crystals of sodium chloride. This is the common salt you put on your chips.

The sodium and chlorine have combined chemically to form a new compound with new properties. Each sodium atom has lost an electron to become a positive sodium ion (Na^+). Each chlorine atom has gained an electron to form a negative chlorine ion (Cl^-). These have joined together to form an ionic compound, sodium chloride. The oppositely charged particles are attracted to form an ionic bond. We can write equations for this:

sodium + chlorine → sodium chloride

$$2Na + Cl_2 \rightarrow 2NaCl$$

The balanced chemical equation represents the simplest proportion of the different ions present in a salt crystal – in this case, 1 to 1. The ions do not actually pair up in this way. They stack up in a giant structure called an ionic lattice. This regular stacking pattern gives rise to the typical cubic shape of salt crystals.

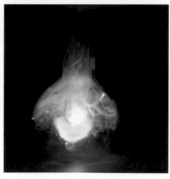
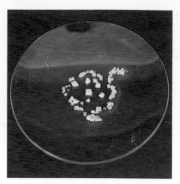

▲ Reactive sodium burns in toxic chlorine …   ▲ … to form common salt.

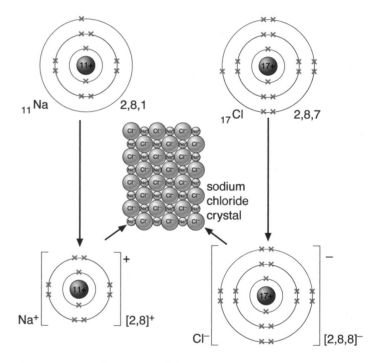

$_{11}Na$ 2,8,1

$_{17}Cl$ 2,8,7

sodium chloride crystal

Na^+ $[2,8]^+$

Cl^- $[2,8,8]^-$

▲ Common salt forms an ionic lattice.

> ### Questions
>
> **a** Calcium loses 2 electrons to form a 2+ ion. Why is calcium chloride $CaCl_2$?
>
> **b** Oxygen gains 2 electrons to form a 2– ion. Why is sodium oxide Na_2O?

The tight grip of the ionic bond

In an ionic compound, every ion is held in place by strong electrostatic forces from its oppositely-charged neighbours. These strong forces are called ionic bonds. Because of these strong bonds, ionic compounds:

- have high melting points and high boiling points

- are quite hard but brittle so they shatter easily

- do not conduct electricity when solid because the charged particles are held tightly in place.

> ### Question
>
> **c** Ions are arranged + against – in an ionic lattice. What would happen if a large force pushed a layer so that + ions were against + ions? (Why are ionic materials brittle?)

Ionic compounds do conduct electricity if the ions are freed from the lattice by melting or dissolving in water. The charged particles themselves carry the current.

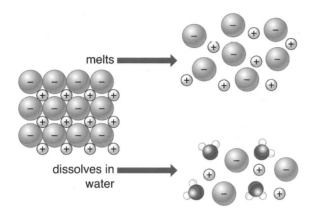

▲ Ionic compounds will only conduct electricity if their ions are freed up by melting or dissolving.

A more flexible structure: metals sharing electrons

Metal atoms can also stack up in a regular way to make a giant structure. All metal atoms have 'loose electrons' in their outermost electron energy levels. In solid metal these outer level electrons are free to move throughout the material. That is why metals conduct electricity.

This 'cloud' of electrons also binds the structure together in a strong but flexible way, allowing the atoms to slide over one another without breaking the material if a force is applied. That is why metals can be beaten or stretched into shape. The electrons are also responsible for heat conduction.

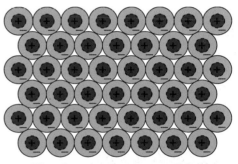

The shared outer electrons bind the metallic lattice together.

▲ Metal atoms stack up in a regular way.

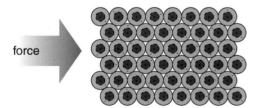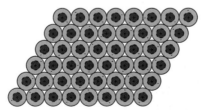

The layers of atoms can slide over one another.

The layers 'stick' back together in the new shape.

Questions

d How could you tell a metal from an ionic compound with a hammer?
e How could you tell a metal from an ionic compound with a battery, a bulb and some wire?

Key points

- Atoms that lose or gain electrons easily are very reactive.
- When atoms react they form ions that have a full outer energy level. This makes the ions very stable.
- Metals lose electrons to form positive ions and non-metals gain electrons to form negative ions.
- These oppositely charged ions are attracted together to form strong ionic bonds.

No electrons to spare?

Non-metals cannot form ionic compounds on their own. Their atoms need to take electrons to get a full energy level, but without metal atoms where can they get them? They need a different way to reach that stable 'full outer energy level' state.

The trick is to share. Non-metal atoms can join their outer energy levels together and share one or more pairs of electrons. When they do this, the atoms join together and we can imagine them looking like soap bubbles. These arrangements are called molecules. Because no electrons have been gained or lost, molecules carry no electrical charge. This arrangement can be very stable and the shared electrons form very strong **covalent bonds**.

Question

a Many gases such as hydrogen and oxygen usually exist as molecules (H_2, O_2). What kind of bonding is there within these molecules?

'Give and take'

Sodium chloride is an ionic compound.

Sharing

Water is a covalent compound.

▲ Ions stay separate; covalent bonds make the atoms stick together like bubbles.

Choosing the right model

Atoms and molecules are far too small to see so we can't draw accurate pictures of them. Instead we use models and symbols to help us visualise what is going on in chemical reactions. As we often want to focus on different aspects of the reaction we use different models for different purposes. Don't let this confuse you. Just use the best model to show what you want it to show.

- Molecular models help to visualise the shape of molecules.
- Electron energy level diagrams show how covalent bonds are formed. These are often simplified to 'dot and cross' diagrams.
- Structural formulae show covalent bonds clearly.
- Simple formulae show the atoms involved.

Examples of each of these are shown on the next page.

The elements that won't go out alone

Chlorine atoms are just one electron short of a full outer energy level, so they need to share just one pair of electrons to form a Cl_2 molecule. This is a single covalent bond. You can show this in an electron energy level diagram. You draw the outer energy levels of the two atoms slightly overlapping. You then arrange the electrons in pairs around the energy level circles. Use dots for electrons on one, and crosses for the electrons on the other. You will end up with one dot and one cross in the 'overlap' where the two atoms have joined together. This is the shared pair of electrons that makes the covalent bond. The 2 electrons are exactly the same – using a dot and a cross just helps you to see how the shared pair have formed.

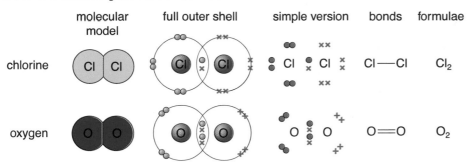

▲ Some different ways to show covalent bonds.

The oxygen in the air also forms a molecule (O_2). Oxygen atoms are 2 electrons short of a full energy level, so they share 2 electron pairs to make a double covalent bond in the O_2 molecule.

Compounds made through sharing

Atoms of different non-metallic elements can also join together in this way to make **covalent compounds**. Chlorine can form a single covalent bond with hydrogen to make hydrogen chloride (HCl). Oxygen needs to share 2 electrons and so can join with 2 hydrogen atoms to make hydrogen oxide – water (H_2O).

Nitrogen is 3 electrons short of a full outer energy level. It can share a pair of electrons with each of 3 hydrogen atoms to make an ammonia molecule (NH_3). Carbon is 4 electrons short of a full outer energy level. It can share a pair of electrons with each of 4 hydrogen atoms to make a methane molecule (CH_4).

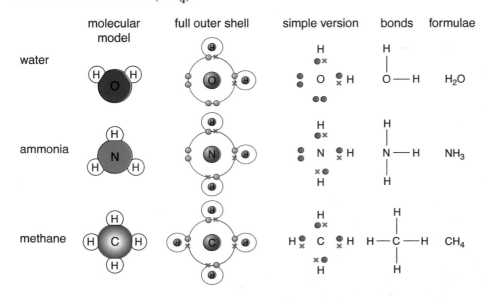

Questions

b Draw an electronic structure diagram for an H_2 molecule. (Remember, the first energy level can only take 2 electrons.)

c Draw an electronic structure diagram for an HCl molecule.

d Carbon can also combine with 2 oxygen atoms forming 2 double covalent bonds. Draw an electronic structure diagram for a CO_2 molecule.

Key points

- Molecules form from non-metal atoms when pairs of electrons are shared between them.
- The bonds holding the atoms together are called covalent bonds.
- Some elements form molecules on their own. Most molecules are compounds.
- We use different models to show covalent bonds for different purposes.

Varieties of bonding

To melt or boil a substance, you need to put enough energy in to break the bonds between the particles. Why is this easy to do for some covalent materials – yet so hard to do for others?

Small molecules

The bonds within covalent molecules are very strong but the particles have no electric charge. This means that covalent materials do not conduct electricity. It also means that there are only very weak forces *between* the particles – **intermolecular forces**. Heating overcomes the forces between the molecules – the covalent bonds remain intact. When water boils, for example, the gas formed is steam and it can condense back into water – maybe as droplets on a cold window. The molecules do not break up into atoms of hydrogen and oxygen.

Because the intermolecular forces are weak, covalent compounds have low melting and boiling points. Small molecules (such as carbon dioxide or methane) form gases at room temperature. Larger molecules may be liquids (such as the hydrocarbons in petrol) or soft solids (such as wax or iodine).

Questions

a Sulfur reacts with oxygen to form small sulfur dioxide (SO_2) molecules. Is this likely to be a solid, liquid or gas?

b Water (H_2O) has small molecules. What is unusual about it?

The macromolecules of carbon

Some non-metals can make giant structures by sharing electrons. Carbon can form diamond and graphite in this way. In these giant structures, every atom is joined to its neighbour by a strong covalent bond. These macromolecular materials have high melting and boiling points, as it takes a lot of energy to break the covalent bonds. They also form hard and strong solids at room temperature.

Diamonds and graphite are both **macromolecules** containing only carbon atoms. Their properties are very different: you wouldn't draw a graph with a diamond, or put graphite in a wedding ring, would you? How can we explain the dramatic difference? Here's a clue: properties such as hardness, as we noted above, are often linked to the way atoms are bonded.

In a diamond, each carbon atom has all four of its outer electrons paired up in covalent bonds with a neighbouring atom. This makes a rigid and hard three-dimensional structure. As there are no 'spare' electrons, diamond does not conduct electricity.

▲ Many jewels, like this amethyst, are giant molecules.

Question

c Silicon and oxygen form silicon dioxide (silica), which has a giant covalent structure. Suggest the likely properties of silica.

In graphite, the carbon atoms are arranged in two-dimensional sheets, with each atom joined to just three neighbours. This leaves one electron 'unpaired' in each atom. These 'loose' electrons can be made to move, so graphite does conduct electricity. Also, there are only weak forces between the carbon sheets, so graphite as a whole is a weak solid as the sheets can slide over one another.

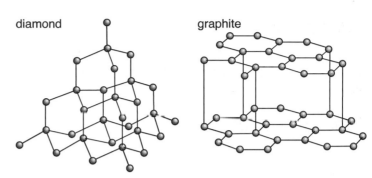

diamond   graphite

▲ Same atoms, but different structures give very different properties.

▲ A mineral to write with.

Questions

d Diamond is the hardest substance known. What does that tell you about carbon–carbon covalent bonds?
e Pencil 'lead' is made from graphite. Why does it leave a trail of carbon when rubbed across paper?

This table shows the properties of some common covalent materials.

Chemical	Uses	Formula	Melting point (°C)	Boiling point (°C)
graphite	pencil 'lead'	C	sublimes (solid → gas)	3825
hexane	in some fuels	C_6H_{14}	−95	67
hydrogen	rocket fuel	H_2	−259	−253
methane	natural gas	CH_4	−182	−162
silica	sand	SiO_2	1410	2355
stearic acid	to make soap	$C_{17}H_{35}COOH$	69	361

Questions

f Which chemicals have small molecules and which have giant covalent structures? How do you know?
g Describe any broad pattern you can see regarding the melting points of the small covalent chemicals.
h What kind of solid would you expect stearic acid to be?

Key points

- The covalent bonds within molecules are very strong.
- The forces between molecules are very weak.
- Very small molecules are gases or liquids. Larger molecules (like wax) may form soft solids.
- Some non-metals form giant molecules where every bond is a strong covalent bond. These form very hard solids with very high melting points.

Properties of matter

We are beginning to see how the arrangement of atoms and molecules inside a material makes a difference to what we can do with it. We wouldn't make jewellery out of paper, for example. Before we use a material for a particular purpose, we need to know whether it is hard or soft, brittle or able to be bent and shaped. These are the material's **properties**. Different materials have different properties. That's why they are used for different jobs.

Question

a *Why are windows made from glass? (Hint: What is the key property?)*

The big picture

The way that different materials are chemically bonded controls their overall properties:

▲ Metals are great for machinery – but diamonds give the best cutting edge.

Type of material	Structure	Strength /hardness	Melting and boiling points	Electrical conductivity	Example	
ionic	giant lattice of positive and negative ions, held together by strong electrostatic forces (ionic bonds)	hard but brittle	high	not when solid; conduct when molten or if dissolved in water		salt crystals
metallic	giant lattice of positive ions surrounded by loose electrons, held together by strong electrostatic forces	hard and strong; can be bent and shaped	high	good (also conduct heat well)		steel cables and girders
small molecular	simple uncharged molecules with only weak forces between them	soft (when solid)	very low – mostly gases at room temperature	do not conduct		air and clouds
giant covalent structure	all bonds are strong covalent bonds	very hard	very high	do not conduct (except graphite)		diamonds

Question

b *Which type of material would be best to use for*
(i) sandpaper (ii) a cable for winching up heavy loads (iii) a jewelled necklace (iv) electrical cables (v) inflating balloons? Explain your answers.

It's all in the detail

Once you know which type of material to use, you of course need to look at the detail. The table shows a range of properties for some different metals.

Metal	Melting point (°C)	Strength	Electrical conductivity	Resistance to corrosion	Cost (£/tonne)
aluminium	660	low	very high	quite high	800
copper	1080	low	very high	very high	1600
gold	1060	low	the best	the best	80 000
steel	1540	very high	low	moderate	300
tungsten	3370	high	moderate	high	30 000

Questions

c Which metal would you use for
(i) the body panel for a car (ii) the wire filament in an electric light bulb
(iii) electric cables (iv) household water pipes (v) a winching cable?
Explain your answers, giving as many reasons for your choice as possible.
d The cables that are hung on electricity pylons to carry electricity for the National Grid are made of aluminium threaded with steel. Use the table to explain why.

Responding to change

Some materials are useful because their properties change. Wax is a solid at room temperature, but if it gets hot it melts. This is used in sprinkler systems that turn on by themselves if a fire starts.

cool — water supply — wax plug (solid at room temperature)

getting warm — water supply

HOT! — water supply — wax plug melts and is pushed out by the water

▲ A wax plug in a sprinkler system.

Smart materials

Materials that change their properties in useful ways like this are called **smart materials**. Now that scientists understand why materials have different properties, they have started to design fantastic new smart materials. These include special dyes that change colour with temperature, materials that change from solid to liquid at the flick of a switch and flexible plastic transistors for 'roll-up' computers or TVs.

Question

f If you could invent a new smart material, what would you want it to do and how would you use it?

Wax	Melting point (°C)
A	29
B	55
C	109

Question

e The table above shows the melting point of three waxes. Which would be good to use in a sprinkler system and why? Explain why the other two waxes would not be suitable.

Key points

● Many properties of materials are determined by the type of bonds between the atoms and molecules that make them up.
● There are four main types of structure: ionic, metallic, small molecular, and giant covalent (macromolecular).
● Properties such as strength and hardness, melting and boiling points, and electrical conductivity depend on the structure of the material.

Technologies too small to see

What is nanotechnology?

A nanometre is a millionth of a millimetre. **Nanotechnology** is a new science that looks at the behaviour of *very* tiny particles – less than one ten-thousandth of a millimetre across. These **nanoparticles** have many potential uses. Some people think that they pose threats to our planet, too. See what you think …

▲ Who's not using nanotechnology?

Tiny particles offer better protection…

Sunblocks use titanium oxide particles to block the harmful ultraviolet (UV) radiation from the Sun. Traditional sunblocks use 'big' particles that appear white on the skin. But you can now buy sunblocks with nanoparticles of titanium oxide. The particles are so small that they don't appear white on your skin. They also cover the skin more evenly, giving better protection from UV radiation.

> **Question**
>
> a A standard titanium dioxide particle has 100 times the diameter of a nanoparticle. How does its volume compare to the nanoparticle volume?

…and make reactions go faster

It's a general rule in chemistry that the smaller the particles, the faster the reaction. Nanoparticles are being developed to speed up important chemical reactions.

- Cars are fitted with catalytic converters to reduce exhaust pollution. New nanoparticle catalysts could make cars and lorries pollution-free.
- Similar chemicals could be used to filter and clean the air, giving us all a healthier environment.
- New efficient nanoparticle fuel cells are being developed to power a range of devices – from cars to your mobile phone.

▲ Nanoparticles for a cleaner, healthier future?

Nano-imprinted chips

Scientists are busy working on the next generation of computers. Some will be polymer, not silicon-based, 'printed' using special nanoparticle inks to make flexible microchips. Others will use nanoparticle tubes of carbon, like rolled up sheets of graphite, to tap into a whole range of new effects, only seen on this tiny scale.

Maybe one day we will have truly intelligent computers. Then we could develop really useful robots. Maybe your children will be taught by one.

> **Question**
>
> b Make a list of potential advantages and disadvantages of truly intelligent robots.

Microscopic machines

Why stop with large robots? The next step is nanoengineering, developing machines that work down close to the size of molecules. These could be developed to:

- be injected into the human body to search out and destroy cancer cells, or perform minor surgery
- be released into the environment to clean up environmental disasters such as wrecked oil tankers
- be sent out to the Moon and planets to explore and report back their discoveries.

You would need many millions and millions of these tiny nanobots for them to have any effect. One 'sci-fi' solution is to make them build themselves. An army of self-replicating nanobots could be set to work on any task we like – from cleaning the house to mining the Moon. But before we rush into using nanotechnology, we need to be sure it is safe.

There may be problems ahead. What if you breathed in super-reacting nanoparticles? What would they do to your body chemistry? How can we be sure that they are safe?

And nanomachines could get out of control. What if they went wrong? What if they … evolved? Some people are scared that they could reduce our green and blue planet to nothing but a 'grey goo' of featureless, lifeless dust.

Recently members of the public were invited to look at the issues. This 'citizens' jury' approved the use of nanotechnologies in healthcare and renewable energy, where they saw potential benefits. But they also called for better labelling and safety testing of manufactured nanoparticles, which some scientists regard as potentially toxic. Above all, they wanted the public to have a greater say in the direction of research.

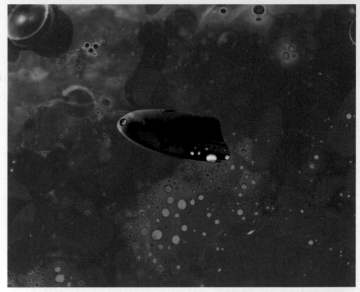

▲ Nanobots in the human body – science fiction or science future?

Questions

c Scientists researching nanoparticles have to keep them in solution or work with extractor hoods to stop them escaping. Why is that?

d What tasks would you give to a group of friendly nanobots?

Question

e Is it scaremongering to talk about such 'worst case scenarios' or is it a sensible precaution to consider the dangers before science progresses too far? What do you think?

Key points

- Nanotechnology is opening up vast new areas for scientific and technological development.
- As with all major advances it has the potential for great good or great harm.
- Which way it goes depends on how we humans choose to use it.

The power of measurement

Modern scientists are now convinced that atoms, though we cannot see them, are the basic particles in matter. Many of the substances we know are compounds made up of atoms of different elements joined together in simple ratios to one another. Such materials are made up of billions of identical atoms or molecules.

In this section we look at just how powerful this insight can be. It can help us to improve the quality of our lives and make the best use of the limited resources available to us on Earth.

▲ Modern scientific instruments rely on computers.

Do you match up to John Dalton?

The proof of the atomic theory is largely due to the work of John Dalton. He made painstaking measurements of the mass of each element in simple compounds. This enabled him to spot the simple ratios that indicated the make-up of the molecules.

Dalton used carefully engineered brass instruments in his detailed work to find the mass of each element in simple compounds such as copper oxide. He:

- tested his balances to make sure they were accurate and reliable
- designed his experiments carefully to make sure they were 'fair tests'
- performed his experiments as carefully as he could to try to avoid errors
- repeated his experiments many times and averaged out his results
- recorded his results clearly so that he didn't get them muddled later.

How do you match up to Dalton, when you do your experiments?

▶ Dalton's instruments were more basic, but very precise for their time.

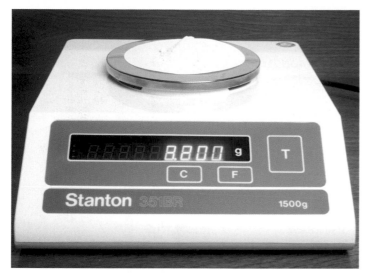

▲ How precise are the balances you use in your laboratory?

Measurement at the heart of it all

Reliable measurement lies at the heart of modern science, and measurement is also important in industry. Nowadays computers handle much of the routine work of making measurements. How much iron will you get from a tonne of iron ore? How much cement from a tonne of limestone? These are vital questions if you want your business to make money. Thanks to the work of Dalton and the scientists that followed him, you can work out the answers to problems like this.

You can also use Dalton's ideas to help the environment. Governments that have signed up to the Kyoto agreement on climate change have to reduce the amount of carbon dioxide produced by their power stations and factories. Carbon dioxide is a gas that just goes up the chimney. How do you measure how much you are making? Actually it's quite easy to calculate it from the amount of fuel you burn. We are going to build on our knowledge of the atom to see how.

▲ How much iron will you get from this shipload of iron ore?

▲ Modern power stations do not belch acid smoke – but they still give off lots of carbon dioxide.

Think about what you will find out in this section

Why different atoms have different masses.

How to work out how much of each reactant is needed in a chemical reaction.

How to work out the yield of a particular product.

How these ideas are used in industry.

One element: two important numbers

All of the elements in the periodic table come with two numbers. The smaller number is the **atomic number** (Z) – the number of protons in the nucleus. This tells you which element it is.

The second, larger number is called the **mass number** (A) and is explained below. You need to understand when and how to use each number. That's what you will learn in this section.

Count the particles

Atoms are so very small that their mass is tiny. A carbon atom has a mass of 2×10^{-23} g, for example. It would be awkward to use these numbers all the time. Instead we just compare the mass of different atoms to each other. Hydrogen, the smallest atom, is 1 on this scale. A helium atom has four times the mass of a hydrogen atom so it is 4. A carbon atom has 12 times the mass of a hydrogen atom so it is 12, and so on. But why are they like this?

The particles that matter

Atoms are made from tiny negative electrons whizzing around a positive nucleus that contains protons and neutrons. As well as having different electrical charges, these different particles have different masses. As with atoms, they are too small to usefully measure in grams, so they can be simply compared to the mass of a hydrogen atom. In this model, protons and neutrons both have a mass of 1, but electrons are so small that their mass is usually ignored. Nearly all of the mass of an atom is in the nucleus.

The mass number of an atom is found by adding the number of protons and neutrons together, ignoring the far lighter electrons. The periodic table tells you the number of protons and the mass number of each element. You can easily work out the number of neutrons: just take the atomic number away from the mass number.

The mass scale: based on carbon

Sometimes more accurate measures of the mass of the atoms of different elements are required. In this case the masses of the atoms are compared experimentally to one-twelfth of the mass of the carbon-12 (^{12}C) atom, which has six protons and six neutrons in its nucleus. The result is called the **relative atomic mass (A$_r$)**. That of hydrogen, for example, is 1.008 when measured in this way.

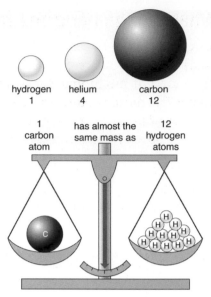

▲ Some common mass numbers.

Particle	Mass	Charge
proton	1	+1 (positive)
neutron	1	0 (neutral)
electron	very small	−1 (negative)

$$\text{mass number } (A) \rightarrow 12$$
$$\text{atomic number } (Z) \rightarrow 6 \quad \text{C}$$

number of protons   (Z) = 6
number of neutrons   (A − Z) = 6

▲ The atomic number and mass number of carbon.

Questions

a If a hydrogen atom and a proton have the same mass, what can you deduce about the sub-atomic structure of a hydrogen atom?

b Work out the number of neutrons in each of these atoms: lithium (Z = 3, A = 7), oxygen (Z = 8, A = 16), argon (Z = 18, A = 40), iron (Z = 26, A = 56).

c How many electrons would be whizzing around each of these atoms?

What are the neutrons for?

A nucleus full of nothing but positively charged protons would be very unstable. The positive charges would repel one another. Neutrons act like a kind of sub-atomic packaging that helps to keep the nucleus stable. The more protons there are the more neutrons you need. The number of neutrons present can vary, even in the same element. There can be slightly lighter and slightly heavier versions of the same element.

Most elements have a relative atomic mass that is very close to their simple atomic mass number. But chlorine has a relative atomic mass of 35.5. What does this mean? Is there 'half a neutron' in its nucleus? Of course not.

The answer is that it is possible to have different versions of the same element which have different numbers of neutrons in the nucleus. Chlorine has two common versions: ^{35}Cl and ^{37}Cl. These have the same atomic number ($Z = 17$) and so are the same element, with the same chemical properties. But one has 18 neutrons in its nucleus, the other has 20. Variants like this are called **isotopes**.

A mixture of isotopes

Most elements have one isotope that is the most common. The relative atomic mass for chlorine is the average of the two isotopes as found in nature. As ^{35}Cl is the more common of the two, the mass number is nearer 35 than 37.

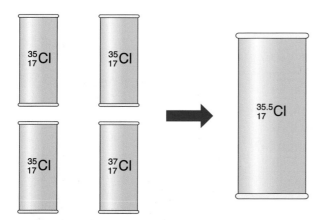

▲ Why chlorine has the relative atomic mass 35.5.

Questions

d Bromine has two isotopes ^{79}Br and ^{81}Br but bromine is usually given as ^{80}Br. What does this tell you about the relative amounts of each isotope that are found naturally?

e How can you tell that rubidium ($A_r = 85.5$) has more than one common isotope?

Key points

- Compared to the mass of a hydrogen atom, protons and neutrons have a mass of 1. The mass of an electron is so small by comparison that it is usually ignored.
- More accurate measurements of relative atomic mass are made by comparison to carbon-12.
- Atoms with the same number of protons are of the same element, but the number of neutrons can vary.
- Atoms of the same element with different numbers of neutrons are called isotopes.

Formula mass

Molecules and other compounds are also very small, so it makes sense to compare their masses on the same scale as atoms. If you know the formula of a compound, you can find its **formula mass** (M_r) by adding up the relative atomic masses of all the atoms that make it up.

For example:

Water: formula H_2O A_r for H = 1, O = 16

formula mass (M_r) = (2 × 1) + 16 = **18**

$$O + H + H \rightarrow \overset{H \quad H}{O}$$

$$16 + 1 + 1 = 18$$

Methane: formula CH_4 A_r for H = 1, C = 12

formula mass (M_r) = 12 + (4 × 1) = **16**

$$C + H H H H \rightarrow \overset{H \quad H}{\underset{H \quad H}{C}}$$

$$12 + (4 × 1) = 16$$

> ### Question
>
> **a** Using the A_r values given, calculate the formula mass for
>   (i) carbon dioxide, CO_2 (C = 12, O = 16)
>   (ii) calcium carbonate, $CaCO_3$ (Ca = 40, C = 12, O = 16)
>   (iii) iron(II) sulfide, FeS (Fe = 56, S = 32)
>   (iv) aluminium oxide, Al_2O_3 (Al = 27, O = 16)
>   (v) magnesium carbonate, $MgCO_3$ (Mg = 24, C = 12, O = 16).

Finding the right formula

How do we know how many of each type of atom join together in a compound? How do we know that it is CO_2 and not C_2O, for example? The answer is by careful experiment. Now that you know about mass numbers, you could do it yourself. Look what happens when magnesium burns in air, for example:

magnesium + oxygen → magnesium oxide

You know that magnesium oxide is made from magnesium and oxygen atoms. But how many of each are there? To take some simple examples, it could be Mg_2O, or MgO or MgO_2. If you know the relative atomic masses of magnesium (24) and oxygen (16), you can work out the percentages by mass of the elements in the compound. Possible results for magnesium oxide are:

▲ What's the formula of magnesium oxide?

Mg₂O

magnesium:oxygen ratio would be $(2 \times 24) = \mathbf{48:16}$
or 75% magnesium, 25% oxygen

MgO

magnesium:oxygen ratio would be $\mathbf{24:16}$
or 60% magnesium, 40% oxygen

MgO₂

magnesium:oxygen ratio would be $\mathbf{24}:(2 \times 16) = \mathbf{32}$
or 43% magnesium, 57% oxygen.

To find which of these is correct, you need to experiment.

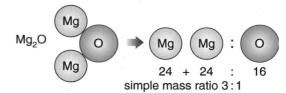

The answer's in the data

In a class experiment, students found the mass of some magnesium, burnt it in a crucible, and then found the mass of the oxide. Here are some results.

	John	Aisha	Kylie	Michelle	Stephan
Mass of magnesium (g)	2.4	3	3.6	2.9	3.2
Mass of oxide (g)	4.0	5	5.6	5	4.0
Mass of oxygen (g)	1.6		2.0	2.1	0.8
% Mg	60		64		80
% O	40			42	

Questions

b Use the figures above to complete the table. Whose results are anomalous (do not fit the pattern)? Suggest a reason for this very different result.

c Ignoring the anomalous result, what is the most likely formula of magnesium oxide?

d Kylie and Michelle's results do not give the exact percentages (but are very close). Suggest a reason for this slight variation.

e In another class, the teacher did one demonstration of the experiment. Why are her results less reliable than the class set?

It is thanks to experiments like this, repeated over and over again in laboratories all over the world, that we know the formulae of chemical compounds. Formulae found by experiment like this are called **empirical formulae**.

Key points

- The relative mass of a compound is found by adding up the relative atomic masses of its atoms.
- You can use figures like this to find the percentage of an element in a compound.
- Empirical formulae are found by experiment.

Nothing gained, nothing lost...

Balanced chemical equations

During chemical reactions, the atoms are simply rearranged from reactant to product. So if you write a symbolic chemical equation you must make sure that there are the same number of each type of atom on each side of the equation.

Not all equations balance straight away. For example, magnesium oxide reacts with hydrochloric acid to give magnesium chloride and water.

Step 1: write out the word equation:

$$\text{magnesium oxide} + \text{hydrochloric acid} \rightarrow \text{magnesium chloride} + \text{water}$$

Step 2: put in the formulae. Count up the atoms on each side to see if they balance:

$$MgO + HCl \rightarrow MgCl_2 + H_2O \qquad Mg\ 1 \rightarrow 1,\ O\ 1 \rightarrow 1$$

… these are OK but …

Step 3: … H 1 $\rightarrow$ 2, Cl 1 $\rightarrow$ 2. This does not balance, but it is easy to fix. You need two lots of hydrochloric acid on the left-hand side, so just put a large 2 in front of the HCl.

$$MgO + 2HCl \rightarrow MgCl_2 + H_2O$$

Now it balances. Some reactions are more complex and may need more 'steps'. Work logically and you will achieve a balance eventually. If necessary, you can check your result in a reference book. Either way, a balanced equation is a powerful tool. You can use it to calculate how much of the reactants you need to make the amount of product you require.

Follow the recipe – and stand well back!

The thermit reaction is a spectacular way to make molten iron. Give a mixture of aluminium powder and iron oxide a kick-start of energy and stand well back. The more reactive aluminium displaces the less reactive iron from its compound in a violent exothermic reaction.

You have to get the proportions right or it doesn't work properly. How can you be sure? Rely on trial and error, and you may disappoint your audience! Knowing the relative atomic and formula masses is vital.

First you need the balanced chemical equation.

$$2Al + Fe_2O_3 \rightarrow 2Fe + Al_2O_3$$

Then you need the relative atomic masses to work out the formula mass.

Al = 27: you need two atoms of Al so the total is 54.
Fe = 56, O = 16: so the M_r of Fe_2O_3 is $2 \times 56 + 3 \times 16 = 112 + 48 = 160$.

So in this reaction, the ratio of the relative masses of the reactants involved, $Al:Fe_2O_3$, is 54:160 or approximately 25% aluminium to 75% iron oxide.

▲ The perfect thermit reaction.

a Why do you get two atoms of iron from every iron oxide 'particle'?
b This reaction produces molten iron. Why is that?
c Work out the formula mass of aluminium oxide. Add this to the combined masses of the two iron atoms. Does it match the sum of the reactants?

Waste costs money

In industry calculations like this are very important. If you get the proportions of your reactants wrong you will end up wasting raw materials. And waste costs money. Working out your reacting masses carefully could make the difference between a healthy profit and going bust.

The refining furnace

Lead, tin and zinc can all be made by carbon reduction in a blast furnace. The metal oxide ore reacts with coke. Coke is made by heating coal. It is almost pure carbon. How many tonnes of ore would you need per tonne of coke? Let's look at the zinc reaction, and then you can work out the others.

The relative atomic mass of zinc is 65, carbon is 12, oxygen is 16. The formula mass of ZnO is therefore $65 + 16 = 81$.
The balanced equation is:

$$2ZnO + C \rightarrow 2Zn + CO_2$$

So in this case, you need $2 \times 81 = 162$ reacting masses of ore for every 12 of carbon.

Turning that ratio into actual masses, you would need 162 tonnes of ore for every 12 tonnes of coke, or $162 \div 12 = 13.5$ tonnes of ore per tonne of coke.

Here are the relative atomic masses and balanced chemical equations for the other metals:

lead   Pb = 207 $PbO_2 + C \rightarrow Pb + CO_2$
tin   Sn = 119 $SnO_2 + C \rightarrow Sn + CO_2$

9 g iron oxide   3 g aluminium

▲ Would these mix for the perfect thermit reaction?

▲ Molten metal pours from a furnace.

Key points

- In chemical reactions, the number of atoms of each element remains the same throughout.
- You can calculate the masses of products and reactants from balanced symbolic equations.
- By knowing how much of each reactant is needed, you can ensure that you get the reaction you want.

d Work out the reacting masses for the reaction with lead ore. From this work out how many tonnes of lead ore would be needed to react with 1 tonne of coke. Repeat this for tin ore.
e How many tonnes of carbon dioxide is produced from each tonne of coke? (Why might this be a problem?)

The same reaction – billion of times

When you write an equation, you describe the reaction between one set of particles. The equation is just as true if you double the number of each type of particle present, or treble them, or multiply them by ten, a hundred, a million – or any number you choose. The relative proportions would still remain the same. We use this fact to help us translate the formulae for chemical reactions into quantities of substance we can work with.

In any reaction you see or perform, many billions of particles are involved. Take carbon, for example, with a relative atomic mass of 12. How many atoms are there in 12 g of carbon? This number is very large indeed. It is approximately 6×10^{23} – that is six hundred thousand million million million.

There are exactly the same number of atoms in 56 g of iron (A_r of Fe = 56). The same applies for any atom – and indeed any molecule. Take any chemical formula and add up the relative atomic masses to get the formula mass. That number of grams of the substance will also contain the same number of particles.

Let's look at making sulfur dioxide (S = 32, O = 16):

$$S + O_2 \rightarrow SO_2$$

So to scale up, for example:

$$25S + 25O_2 \rightarrow 25SO_2$$
$$\text{or} \quad 579S + 579O_2 \rightarrow 579SO_2$$
$$\text{or} \quad 6 \times 10^{23}\,S + 6 \times 10^{23}\,O_2 \rightarrow 6 \times 10^{23}\,SO_2$$
$$\text{or} \quad 32g\,S + 32g\,O_2 \rightarrow 64g\,SO_2$$

A mole of particles – always the same number

This number (6×10^{23}) is used as a standard number of particles. It is based on the number of particles in exactly 12 g of carbon-12 and is called the **mole**. You can have a mole of atoms, molecules, ions – even a mole of electrons. Whatever it is, if you have 6×10^{23} of them, you have a mole. It is really quite a simple idea: you have 1 mole if you have the relative atomic mass (or formula mass) in grams.

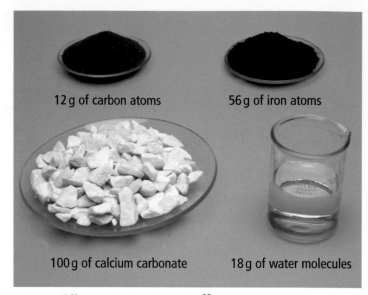

12 g of carbon atoms   56 g of iron atoms

100 g of calcium carbonate   18 g of water molecules

▲ Some different ways to get 6×10^{23} particles.

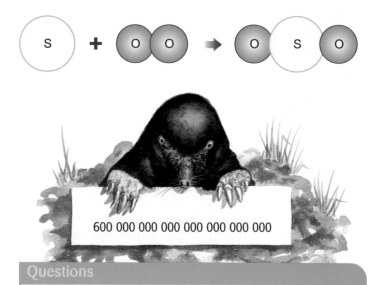

600 000 000 000 000 000 000 000

Questions

a How many moles are there in
   (i) 18 g of water (ii) 36 g of water (iii) 32 g of sulfur dioxide?
b How many grams would you need to have 0.1 mole of carbon?

Atoms or molecules? Keep a clear head

In order to know the mass needed for 1 mole of something, you need to be clear about the particles involved. In many cases this is straightforward.

1 mole of magnesium atoms: A_r of Mg = 24, so 1 mole of Mg = 24 g.

1 mole of carbon dioxide molecules: M_r of CO_2 = 12 + 2 × 16 = 44, so 1 mole of CO_2 = 44 g.

But what about 1 mole of oxygen? You need to decide if you are referring to oxygen atoms or oxygen molecules.

A_r of O = 16, so 1 mole of oxygen atoms = 16 g.

M_r of O_2 = 32, so 1 mole of oxygen molecules = 32 g.

In a gas, volume's the key

It's easy to find a mole of a solid or pure liquid – you just weigh out the right amount. But what about a gas? It's not easy to find the mass of a gas. With gases, it's much easier to find the volume. How does that link up with the idea of a mole?

Scientists studying gases discovered something interesting about 200 years ago. If you have the same volume of gas, you have the same number of particles, whatever the gas (as long as the pressure and temperature are the same). So $1 dm^3$ of hydrogen contains as many particles as $1 dm^3$ of chlorine, or $1 dm^3$ of methane, and so on.

This makes it easy to use moles when the substance you are looking at is a gas. At room temperature and atmospheric pressure, 1 mole of any gas has a volume of $24 dm^3$.

Questions

c What mass would give you 1 mole of
(i) sulfur atoms (S)
(ii) sulfur molecules (S_8)?
d How many moles are there in
(i) 48 g of oxygen atoms (O)
(ii) 48 g of oxygen molecules (O_2)
(iii) 48 g of ozone molecules (O_3)?

▲ It's not easy to find the mass of a gas.

oxygen   chlorine   nitrogen dioxide

▲ There are the same number of molecules in each jar.

Questions

e At atmospheric pressure and room temperature, how many moles are there in
(i) $6 dm^3$ of sulfur dioxide, SO_2 (ii) $24 dm^3$ of methane, CH_4, (iii) $48 dm^3$ of nitrogen dioxide, NO_2?
f What would be the volume of 1 g of hydrogen gas (H_2) at atmospheric pressure and room temperature?

Key points

- If you have 1 mole of a substance you have $6 × 10^{23}$ particles of it.
- You have 1 mole of atoms if you have the relative atomic mass in grams, 1 mole of molecules if you have the formula mass in grams, and so on.
- One mole of any gas has a volume of $24 dm^3$ at room temperature and pressure.

Choosing what to start with

Suppose you start with a tonne of ore. What is the most metal you could possibly extract from it? Let's look at extracting iron. Both haematite (Fe_2O_3) and pyrites (FeS_2) can be used as iron ore. Say they cost the same. Which would give you the most iron for your money? Once you know the formula mass it's easy to work out the percentage of iron. (A_r values: Fe = 56, O = 16, S = 32.)

Haematite (Fe_2O_3)

Formula mass = $2 \times 56 + 3 \times 16 = 112 + 48 = $ **160**.
Amount of iron = $2 \times 56 = $ **112**.

Percentage of iron = $\dfrac{112}{160} \times 100\% = $ **70%**.

Pyrites (FeS_2)

Formula mass = $56 + 2 \times 32 = 56 + 64 = $ **120**.
Amount of iron = **56**.

Percentage of iron = $\dfrac{56}{120} \times 100\% = $ **47%**.

▲ Haematite, Fe_2O_3.

▲ Iron pyrite, FeS_2.

The best you can get

How much of a product you get from your raw material is called the **yield**. If your haematite iron ore was pure and your process was perfect, you would get 70 tonnes of iron for every 100 tonnes of haematite. This is the **theoretical yield**. In reality you would get less than this because:

- your ore may have contained impurities
- the reaction may not have gone to completion
- some iron may not have separated out from the reaction mixture
- some other unexpected reactions may have occurred.

What if you only got 63 tonnes of iron? You could work out what percentage this was of the theoretical yield. In this case the **percentage yield** is:

$$\frac{\text{actual yield}}{\text{theoretical yield}} = \frac{63}{70} \times 100\% = \mathbf{90\%}.$$

Turning down the heat

Carbon dioxide is a greenhouse gas. The way we burn fossil fuels is thought to be responsible for global warming. But how can we tell how much is produced?

Questions

a Pyrites is not often used as iron ore for obvious reasons. Another ore is magnetite (Fe_3O_4).
(i) What is the percentage of iron in magnetite?
(ii) Which would you buy if they cost the same?
(iii) Which would you buy if magnetite was 10% more expensive? Explain your answer.

b What would be the percentage yield if you got 66.5 tonnes of iron from 100 tonnes of haematite?

Question

c Before reading on, have a guess at the amount of carbon dioxide that is produced by using 1 dm³ (700 g) of petrol in a car engine – is it about 2 kg, 200 g, 20 g or 2 g?

▲ Flying you to New York produces over 3 tonnes of carbon dioxide per passenger. That's the mass of an elephant.

Petrol is made of hydrocarbons similar to octane, C_8H_{18}. When it burns:

octane + oxygen → carbon dioxide + water

The formula mass of octane is $(8 \times 12) + (18 \times 1) = 114$.

The formula mass of carbon dioxide (CO_2) is $12 + (2 \times 16) = 44$.

When a molecule of octane burns it produces 8 molecules of carbon dioxide, one for each of its carbon atoms. That makes $8 \times 44 = 352$.

So the mass ratio is

(not balanced) $C_8H_{18} \rightarrow 8CO_2$
$114:352$ that's more than $1:3$.

So $1\,dm^3$ of petrol produces $3 \times 700\,g = 2.1\,kg$ of CO_2.

Even breathing warms the planet

You produce carbon dioxide when you respire. Inside your cells, glucose reacts with the oxygen you breathe to get the energy for life. Every molecule of glucose ($C_6H_{12}O_6$) that reacts gives 6 molecules of carbon dioxide.

So the mass ratio is

(not balanced) $C_6H_{12}O_6 \rightarrow 6CO_2$
$180:264$ that's nearly $2:3$.

Many snack bars are 50% glucose or similar carbohydrates. So that 40 g bar you ate at break (containing 20 g glucose) could have you breathing out 30 g of carbon dioxide if it reacted completely.

Questions

d Travelling produces a lot of carbon dioxide. Work out how much is produced by these London to Oxford trips. How much per person?
(i) Sheela and Austin drove in their economy car. They used 5 dm³ of petrol.
(ii) Otis drove in his big 4x4. He used 12 dm³ of petrol.
(iii) Oliver took the bus. It used 50 dm³ of petrol, but was carrying 50 passengers.
e 'Enercrisp' bars contain 30 g of glucose. How much carbon dioxide could you breathe out if you eat one? Why might the actual 'percentage yield' be slightly lower than this?

Key points

- The maximum possible amount of a product that could be produced in a reaction is called the theoretical yield.
- In the real world, the actual yield is often lower. The ratio of the actual yield to the theoretical amount is called the percentage yield.

Reactions that don't go all the way

Forwards and backwards

So far you have looked at chemical reactions as one-way processes:

reactants → products.

For example, if you burn wood you cannot easily get the wood and oxygen back from the products, carbon dioxide and water. But you also know many reversible processes. Water turns to ice if cooled, but melts back to water when heated. This happens in some chemical reactions too.

You should be familiar with blue copper sulfate crystals. Water molecules are chemically bound into the crystal structure. These can be driven off by heating, leaving a white powder called anhydrous copper sulfate.

Now add water to the powder. The water immediately recombines with the copper sulfate and the blue colour reappears. This reversible reaction is a chemical test for the presence of water.

$$\text{hydrated copper sulfate} \xrightleftharpoons[\text{water added}]{\text{water driven off by heat}} \text{anhydrous copper sulfate} + \text{water}$$
$$\text{(blue)} \qquad\qquad\qquad\qquad \text{(white)}$$

Putting the heat on

If you heat ammonium chloride, it turns directly from a solid to a gas (it **sublimes**), only to reappear as a white solid at the cool end of the tube. This is not a simple physical change. Actually, it is another reversible chemical change.

$$\text{ammonium chloride} \xrightleftharpoons[\text{cool}]{\text{heat}} \text{hydrogen chloride} + \text{ammonia}$$
$$\text{(white solid)} \qquad\qquad\qquad \text{(colourless gases)}$$
$$NH_4OH \qquad\qquad\qquad\qquad HCl \quad + \quad NH_3$$

Heating drives the reaction to the right, but the reaction reverses as it cools down.

Forward and back reactions

You can think of reversible reactions in terms of the forward reaction (left to right) and the back reaction (right to left).

$$A + B \underset{\text{back reaction}}{\overset{\text{forward reaction}}{\rightleftharpoons}} C + D$$

Question

b Describe the 'sublimation' of ammonium chloride in terms of the forward and back reactions.

▲ Most chemical reactions are 'one-way'…

▲ … but some are reversible.

Question

a Anhydrous means 'without water'. What do you think hydrated means?

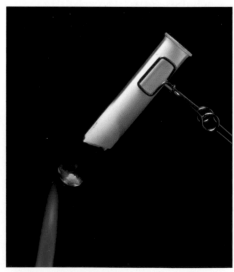

▲ Ammonium chloride breaks up when heated, only to re-form at the cool end of the tube.

If you pass steam over heated iron, the iron is oxidised to iron oxide and the steam reduced to hydrogen. This is the forward reaction. But if the hydrogen produced is now passed back over the hot iron oxide, the reaction is reversed and the oxide is reduced. This is the back reaction.

forward reaction

iron + steam ⇌ iron oxide + hydrogen

back reaction

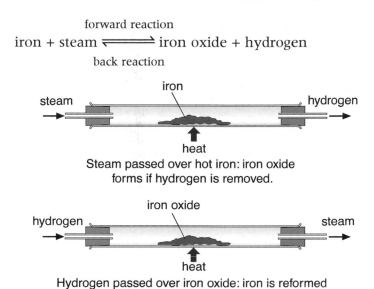

Steam passed over hot iron: iron oxide forms if hydrogen is removed.

Hydrogen passed over iron oxide: iron is reformed if the steam is removed.

All four substances exist in a closed system.

Open and closed systems

This experiment only gets to one end point or the other if one of the products is removed from the reaction site each time. This is called an **open system**. If iron and steam were heated in a **closed system** (such as in a sealed, pressure-proof container), the reaction would stick part-way, with all four molecules present. It would reach **equilibrium** (shown in equations by the symbol ⇌).

$$3Fe + 4H_2O \rightleftharpoons Fe_3O_4 + 4H_2$$

You might think that at equilibrium the reaction has stopped, but that is not the case. Some iron and steam particles are still reacting to give iron oxide and hydrogen, but they are balanced out by other iron oxide and hydrogen particles reacting to give iron and steam again.

Question

c In a big department store, people are constantly travelling up and down the escalators, yet the number of people on each floor remains about the same. Use this as an analogy to explain how chemical equilibrium works.

Key points

- Some chemical reactions are easily reversible.
- In a closed system, the forward and back reactions will reach equilibrium.
- Reversible reactions can be made to go 'one way' if one of the products is removed.

Nitrogen for fertilisers

Plants need nitrogen to grow well. Unfortunately, plants cannot use the vast amount of nitrogen in the atmosphere directly. That's why modern agriculture relies on fertilisers to keep crop yields high.

A century ago, demand for food in Europe was outstripping supply. The growing population in Europe meant the demand for food was high. The only fertilisers available to farmers were materials easily to hand, such as manure. Chemists needed to find a way to 'fix' the atmospheric nitrogen so that plants could use it. The race was on to find a way to feed the people (and get rich in the process).

Chemists help out on the farm

One nitrogen-rich compound that plants can use is the pungent-smelling gas, ammonia (NH_3). Early attempts to make ammonia from nitrogen and hydrogen were unsuccessful. The reaction is reversible and in the early experiments reached equilibrium with less than a 10% yield of ammonia – enough to turn damp litmus blue, but that was all.

$$\text{nitrogen} + \text{hydrogen} \underset{\text{back reaction}}{\overset{\text{forward reaction}}{\rightleftharpoons}} \text{ammonia}$$

$$N_2 + 3H_2 \rightleftharpoons 2NH_3$$

Haber seeks a better way

In 1904, a German chemist called Fritz Haber looked at ways of improving the yield. He found that he got the highest yield at high pressure and low temperature (see section 4.4 of this unit). But the yield was still less than 50%, and the reaction was so slow that it would never be economical. If he tried to make the reaction run faster by raising the temperature, the yield just got smaller and smaller. What could be done?

▲ It needs a good supply of fertiliser to get all this plant growth.

Questions

a Sodium nitrate ($NaNO_3$) is mined in some parts of the world and is used as a fertiliser. Ammonium nitrate (NH_4NO_3) is an artificial fertiliser made from atmospheric nitrogen. Calculate the percentage of nitrogen (by mass) in each. (Use Na = 23, N = 14, H = 1, O = 16.)

b How many moles of hydrogen (H_2) are needed to react completely with 1 mole of nitrogen (N_2)?

It's a tricky problem…

Know the properties – they'll suggest an answer

The problem seemed to have no solution. A catalyst made the reaction run faster but how could the yield be increased? The solution came by thinking 'outside the box'. Everyone had been thinking about the reaction as a closed system, but what if it was opened up? If the ammonia could be removed from the mixture it would stop the back reaction, but how could it be done? Haber looked at the boiling points of the gases and saw the solution.

In the modern Haber process, nitrogen (purified from the air) and hydrogen (from natural gas) are reacted over an iron catalyst. At 450 °C and 200 atmospheres pressure the reaction is fast, but still only gives a 30% yield of ammonia.

Gas	Boiling point (°C)
hydrogen	−252
nitrogen	−183
ammonia	−33

Question

c How could the ammonia be removed? Can you spot the key to the puzzle in this table?

▲ A modern Haber plant.

Question

d You would get a higher percentage yield if you ran the Haber plant at room temperature. Why is it not practical to do this?

Haber's great breakthrough was to cool the mixture of gases. The mixture passes through a condenser, which cools the gases below −33 °C, the boiling point of ammonia. The ammonia liquefies and flows out of the condenser.

The unreacted nitrogen and hydrogen are then 'topped up' with more reactants and sent through the system again in a continuous loop. In this way, all the reactants are eventually converted to ammonia.

The world population continues to grow and is now four times as large as it was in Haber's time. Many people across the world owe their lives to Haber and the great leap forward in food production he helped bring about. He was awarded the Nobel Prize for his work in 1918.

Key points

● Making ammonia for fertilisers from the nitrogen in the air is difficult as the reaction is reversible.
● The Haber process overcomes the problem by liquefying and removing the ammonia.

Cleaning up our act

Making and disposing of plastics and other products we need can harm the environment. Stringent laws now govern industrial processes that could cause pollution. Greater environmental awareness has led to the reclamation of many areas once laid waste by pollution. With a wider view of the issues our chemical industry has 'cleaned up its act'.

What about poorer countries?

We should have no problem maintaining this approach in Britain in the future. But what about the rest of the world? As China and India industrialise they will need to build more and more chemical factories and power stations, using more and more natural resources and creating more and more waste. They are now going through the kind of changes Britain went through 50 to 100 years ago. How can their citizens improve their lives without adding even more pollution? That is the problem of **sustainable development**.

Efficiency and atom economy

One way that better science has improved the environment is through efficiency. In a world of limited resources, we need to make the most of what we have. The idea of yield and yield efficiency (see section 2.5 of this unit) is very important in helping us minimise the resources we use, but it is not the whole story. If we want to reduce waste, we have to look at the reaction itself to see how much useful product is produced compared to the total mass of atoms used. This is called the **atom economy**.

Question

a New housing is often built on 'brownfield' sites (areas once used for industry). Why do you think these sites need careful 'cleaning up'?

China and India are making a great mistake. If their huge populations used the same resource per person as people in the USA or Britain, the whole planet would be destroyed. They should stay as they are.

The people of China and India want the same standard of living as us, and we have no moral right to stop them achieving it. If we want to save the planet we must share out technology and resources with them and help them to develop in a sustainable way.

Question

b Discuss the arguments shown. Who do you agree with?

For example, in the final stage of the blast furnace reaction, iron oxide is reduced by carbon monoxide:

$$Fe_2O_3 + 3CO \rightarrow 2Fe + 3CO_2$$

The total mass of the products is
iron (2×56) + carbon dioxide (3×44) = $112 + 132 = 244$.

The atom economy $= \left(\dfrac{\text{mass of useful product produced}}{\text{total mass of all products}}\right) \times 100\%$

$= \left(\dfrac{112}{244}\right) \times 100 = 46\%$, in this case.

So even if the process is run as efficiently as possible, it still produces more waste than product. In this case the waste is carbon dioxide. If scientists could find a way of making iron that produced less carbon dioxide as a by-product, it would be better for the environment.

Question

c Iron could be obtained from iron pyrites (FeS_2) by roasting it in air then reducing the oxide. The combined reactions could be shown as:
$$FeS_2 + 2O_2 + 2C \rightarrow Fe + 2S + 2CO_2$$
(i) Calculate the atom economy of this reaction for iron (S = 32).
(ii) Why is this reaction not used to make iron?
(iii) Why should we be concerned that carbon dioxide is produced as a waste from the manufacture of iron?

Cutting waste in killing pain

Ibuprofen is a popular pain-killer. Until 1980, only doctors could prescribe it. It was manufactured using a six-stage chemical process, with a low atom economy. Then the law changed and people could buy it without a prescription. As demand increased, research intensified and in the 1990s a new three-stage method was developed with a high atom economy.

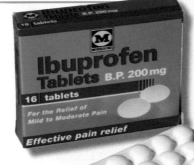

Question

d (i) The formula mass of Ibuprofen is 206. The total waste products in each case add up to 308.5 and 60 for the earlier and later processes respectively. Calculate the atom economy percentages in each case.
(ii) Why do you think the pharmaceutical companies were keen to find a better way to make Ibuprofen?
(iii) What are the wider advantages of the second method?
(iv) Discuss how the concept of atom economy can be of benefit to both business and the wider world.

Key points

- Sustainable development allows us to improve our standard of living without damaging the environment.
- Understanding atom economy helps us to maximise the way we use resources and so minimise our impact on the environment.

Chemical reactions may be very slow – like the slow rusting of an old car, or very fast – like the petrol/air explosions that drive the car's engine. Many, of course, lie somewhere between the two extremes.

Following the reaction

Either way, reaction rates are very important in industry.

- If a reaction runs too quickly, it might get out of control and cause an explosion.
- If it runs too slowly, this will make the process very inefficient and will raise production costs.

Reaction rates need to be controlled, but before you can do that you need to be able to measure them. In practical terms, you need to choose something to measure that indicates the progress of the reaction. Then you keep track of that quantity over a sensible time interval. Examples of useful things to measure are:

- the mass of a reactant used up during a reaction
- the volume of a gas (product) produced during a reaction.

The **rate of reaction** is 'what happens' divided by 'how long it takes', that is:

$$\frac{\text{amount of reactant used}}{\text{time}} \quad \text{or} \quad \frac{\text{amount of product formed}}{\text{time}}$$

Speed it up

Often in industry you are trying to make a reaction go faster. This can make the industrial process more efficient and so help to make more money for your company. In this section you will look at the factors that can affect the rate of reaction and how you can control them. These factors are:

- temperature (Does food cook slower or faster in a hotter oven?)
- surface area (Biscuits don't burn well – but biscuit dust in a factory has been known to cause an explosion.)

▲ Rusting is a relatively slow reaction.

▲ The fuel depot explosion in Hemel Hempstead in 2005 shows what can happen in a fast reaction.

▲ Fine dust can cause an explosion. This grain silo in Blaye, France, was destroyed by a dust explosion in 1997.

- concentration (in liquids) or pressure (in gases). (Bleach is used 'neat' down the drain, but is diluted with water if used in the 'whites' wash.)

You will also find out how catalysts can improve the speeds of reactions without being used up (see section 3.4). This helps the chemical industry run efficiently, but it can also help clean up the environment.

▲ Athens, Greece: traffic pollution can cause this …   ▲ … but catalysts can stop it.

The haze of brown smog you get in big cities is caused by gases from car exhausts. The material emitted contains carbon monoxide and nitrogen oxides which react with the oxygen in the air to form brown nitrogen dioxide.

It is possible to prevent the smog. Before the hot gases get out into the air, pass them across a material that helps them to react. The toxic carbon monoxide is oxidised to carbon dioxide while the nitrogen oxide is reduced to harmless nitrogen:

$$2CO + 2NO \rightarrow 2CO_2 + N_2$$

All new cars are now fitted with 'catalytic converters' by law. Most chemical factories use their own, particular 'catalytic converters' to help their reactions run smoothly and cost-effectively.

Think about what you will find out in this section

What chemical processes can I see going on around me?

How can you measure rates of reaction?

What factors control the rate of a reaction?

How can you speed up or slow down a reaction?

How can science help industry to work efficiently?

Getting things moving

What happens during a chemical reaction? In order to react with each other, particles have to approach one another. Atoms cannot possibly 'change partners' in a chemical reaction unless the particles that contain them come into contact.

That on its own is not enough, though. If it were, most chemical reactions would be almost instantaneous. The particles must also collide with enough energy to break the existing chemical bonds and reform in a new way.

Breaking the chemical bonds that hold a molecule together requires energy. When particles collide, a certain amount of energy becomes available. It may or may not be enough to break the bonds.

Temperature matters

At any given temperature, collisions of many different types will occur. They will range from the maximum energy 'head-on' impacts to gentler glancing blows. The higher the temperature, the higher the average speed of the particles. Faster particles carry more **kinetic energy**. If the temperature is raised, the average speed per particle increases. Collisions that generate enough energy to break the bonds will occur more often, so a larger proportion of the particles have a chance to react. That's why raising the temperature speeds up the reaction.

The minimum energy required for any given reaction is called its **activation energy**. Many reactions need a 'kick-start' of energy to provide this and get the reaction started.

Monitoring the reaction

If you put magnesium ribbon into sulfuric acid it produces hydrogen gas:

$$\text{magnesium} + \text{sulfuric acid} \rightarrow \text{magnesium sulfate} + \text{hydrogen}$$
$$\text{Mg} + \text{H}_2\text{SO}_4 \rightarrow \text{MgSO}_4 + \text{H}_2$$

You can collect this in a gas syringe, and time how long it takes to produce $100\,\text{cm}^3$ of gas, for example. If you run this experiment at different temperatures using a water bath, you can see how this affects the way the hydrogen is produced. You may not be surprised to find that the gas is produced faster when you raise the temperature.

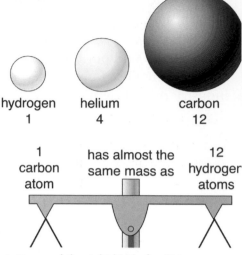

hydrogen   helium   carbon
1   4   12

| 1 carbon atom | has almost the same mass as | 12 hydrogen atoms |

▲ You need the right kind of collision before you get a reaction.

▲ Milk turning sour is a chemical reaction.

Questions

a   *These bottles of milk are the same age. Which do you think was left in a warm room, and which was left in the fridge?*

b   *Why doesn't a Bunsen burner light as soon as you turn on the gas?*

c   *What does a spark provide that sets off this reaction?*

d What factors, other than temperature, will affect how fast the hydrogen is produced?

e What would you have to do to control these other factors, to ensure that they were not affecting the results?

The pace hots up

You could plot a graph of the time it takes to produce the gas against temperature, but it is better to convert the figures to show the rate of production, and plot a graph of that instead.

$$\text{Rate of hydrogen production} = \frac{\text{volume of gas } (cm^3)}{\text{time taken}}$$

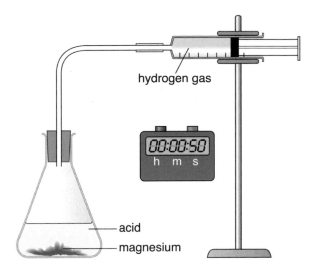

▲ Collecting the hydrogen in a gas syringe.

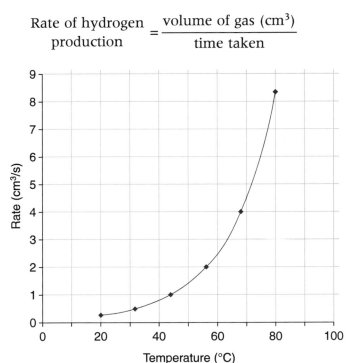

▲ Graph showing the rate of the reaction plotted against temperature.

Question

f From the graph, approximately what rise in temperature doubles the reaction rate?

Looking at a rate/temperature graph you might expect all industrial reactions to be run at as high a temperature as possible. High temperatures are often used but there are limits as the products may themselves break down if heated too much.

Cost is also important. You need a lot of energy to reach very high temperatures, and energy costs money. Industrial chemists are always on the lookout for the most cost-effective ways to run their chemical reactions.

Key points

- For chemicals to react their particles have to collide with enough energy to break the existing bonds.
- The minimum energy for this is called the activation energy for the reaction.
- Increasing the temperature gives particles more energy, so reactions usually go faster.
- To find the rate of a reaction, divide 'what happens' by 'how long it takes'.

Free up some surface

Let's look at some other ways of making collisions happen more often. If one of the reactants is a solid you can do this by making the pieces smaller. Reactions can only take place at the surface of a solid. Breaking up the pieces will dramatically increase the surface area, making available far more opportunities for reactions between molecules.

You've seen this in action on bonfire night. Large lumps of iron don't burn well, but tiny iron filings do. This idea has many uses. If you've ever tried to light a campfire you'll know that it's much easier if you start with small twigs and splinters. You won't be able to set a large log on fire with just a match.

Questions

a   *Why does chewing your food well help you to digest it?*

b   *Flour mills and biscuit factories can easily produce clouds of flammable dust. Why is this a danger? And what could they do to overcome it?*

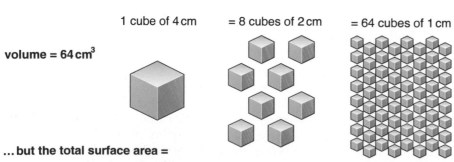

▲ Sparklers are made from tiny iron filings glued onto an iron wire. Iron filings burn because they have such a large surface area.

The heart of the matter

Every time you break up a solid, you expose an extra bit of material to react. If you have the same amount of material in pieces half the diameter, you will have twice the surface area. That will double the reaction rate.

1 cube of 4 cm   = 8 cubes of 2 cm   = 64 cubes of 1 cm

volume = 64 cm^3

... but the total surface area =

$1 \times 6 \times 16 =$ **96 cm^2** $8 \times 6 \times 4 =$ **192 cm^2** $64 \times 6 \times 1 =$ **384 cm^2**

▲ Smaller cubes, larger surface area.

Question

c   *If you broke a 1 cm cube up into 1 mm cubes, by what factor would the surface area increase?*

Chip off the marble block

Marble chips are a natural form of calcium carbonate. They react with hydrochloric acid, giving off carbon dioxide gas.

calcium + hydrochloric → calcium + water + carbon
carbonate   acid   chloride   dioxide

$$CaCO_3 + HCl \rightarrow CaCl_2 + H_2O + CO_2$$

One large lump of marble dropped into a beaker fizzes steadily. But the same amount of marble in powdered form foams up out of the beaker. The acid particles can only collide with the carbonate particles exposed at the surface. Crushing a marble chip like this could increase the surface area a thousand-fold. The reaction rate goes up accordingly.

▲ Which beaker had the powdered marble?

Lost to the air

If this reaction takes place in an open beaker, the carbon dioxide escapes into the air. If this is done on a balance, the total mass will go down as the carbon dioxide is lost. If you measure how much gas is lost in a given time, you can work out the rate of the reaction.

For example, in an experiment using 1 mm marble chips, the mass fell from 50 g to 45 g in 50 seconds.

Mass lost (amount of carbon dioxide produced): 5 g.
Time taken: 50 seconds.

$$\text{Rate} = \frac{\text{amount of carbon dioxide (g)}}{\text{time taken (s)}} = \frac{5}{50} = 0.1 \text{ g/s}$$

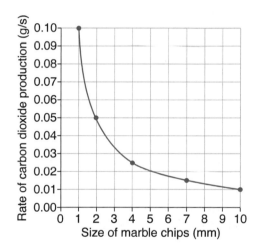

The graph shows that the reaction rate speeds up when you use smaller marble chips.

Questions

d (i) *Using the graph, find the mass loss per second for the 2 mm chips.*
(ii) *How long would it take the mass to drop by 1 g for these chips?*
e *How much faster did the 2 mm marble chips react, compared to the 4 mm chips?*
f *Why is it that delicate carvings on churches seem to suffer more from the effects of acid rain than large blocks made from the same stone?*

Key points

● Reactions involving solids take place at surfaces.
● To make a reaction go faster, increase the available surface area by making the pieces smaller.

Close encounters

Remember that before two particles can react, they must meet. If a substance dissolves in water or another liquid, we can prepare it as a solution to make the molecules accessible to other reactants. We can improve the chance of such meetings in solutions by making them as concentrated as possible. More reactions will occur if more molecules are available.

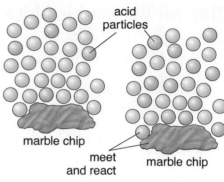

low concentration   higher concentration

The bite of acid

The reaction of magnesium in sulfuric acid (see section 3.1 in this unit) can be used to show how reaction rate changes with concentration. With the acid at low concentrations, the particles are widely spread in the water. The number of collisions between them and the magnesium will be limited. At higher concentrations, the chance of a collision between the acid particles and the magnesium is greatly increased.

Question

a *From this explanation would you expect the reaction rate to rise or fall as you increase the concentration?*

Improving accuracy

Using a gas syringe to collect the hydrogen (see section 3.1) you can follow the way reaction rate changes with concentration. Take readings every few seconds and plot a graph of the volume of hydrogen formed against the time. The line starts off steep and then steadily flattens out as the reaction reaches its **end point** where all the reactants are used up. This is a more accurate way to follow the reaction than taking a single reading, as when you draw the line of best fit, you 'average out' your experimental errors.

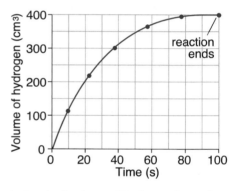

▲ Graph of volume of hydrogen formed against time.

Questions

b *What is the rate of hydrogen production over the first 20 seconds?*
c *How is the rate of reaction shown here changing with time?*
d *What is happening to the concentration of the acid, to give this effect?*
e *Why not use the 'mass loss' method for this? (Hint: compare the formula masses of CO_2 and H_2.)*

Measures of concentration

The concentration of a solution is usually measured in moles per cubic decimetre (mol/dm^3).

The formula mass of sodium hydroxide (NaOH) is $23 + 16 + 1 = 40$. A solution with 40 g of sodium hydroxide in $1 \, dm^3$ of solution would have a concentration of $1 \, mol/dm^3$.

The formula mass of hydrogen chloride (HCl) is $1 + 35.5 = 36.5$. A solution of hydrochloric acid with 36.5 g of the compound dissolved in $1 \, dm^3$ of solution would also have a concentration of $1 \, mol/dm^3$.

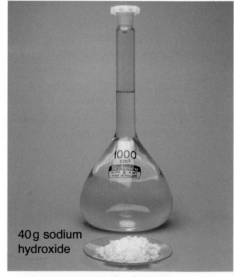

40 g sodium hydroxide

▲ $1 \, dm^3$ of $1 \, mol/dm^3$ solution contains 1 mole of particles.

The same volume of any solution with the same concentration will contain the same number of particles. So for the reaction:

$$NaOH + HCl \rightarrow NaCl + H_2O$$

1 dm³ of **1** mol/dm³ NaOH will exactly neutralise
1 dm³ of **1** mol/dm³ HCl

0.2 dm³ of **3** mol/dm³ NaOH will exactly neutralise
0.2 dm³ of **3** mol/dm³ HCl

Questions

f How many sodium hydroxide 'particles' are there in the flask shown in the photograph on page 124?

g How many dm³ of **0.27** mol/dm³ NaOH will exactly neutralise **0.163** dm³ of **0.27** mol/dm³ HCl? (Hint: you don't need a calculator.)

Gas pressure

The same arguments apply to gases. Pressure takes the place of concentration as an indication of how many molecules are present. All else being equal, the greater the pressure, the greater the number of particles of gas in a given space. If you double the pressure you will have twice as many particles.

The closer the particles are together, the more chance they have of colliding. If you double the pressure, you will double the rate of reaction. Graphs drawn for rate against concentration and rate against pressure look very similar.

Calculations for gases can be even simpler. If you have the same volume of gas, you will have the same number of particles. So to make water:

$$2H_2 \quad + \quad O_2 \quad \rightarrow \quad 2H_2O$$

2 volumes + 1 volume → 2 volumes

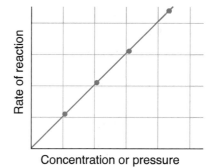

▲ Typical graph of reaction rate against concentration or pressure.

Question

h Water can be electrolysed to give hydrogen and oxygen. If you got 100 cm³ of hydrogen, how much oxygen would you get?

Key points

- For solutions, if you increase the concentration you increase the rate of reaction.
- For gases, if you increase the pressure you increase the rate of reaction.
- Concentrations and pressures can be expressed in units of moles per unit volume (mol/dm³).

Everlasting activity!

We are going to look at one final way of speeding up reactions. Ever helped others get things done just by being there with them and smiling? Some chemicals have the same effect on reactions. Just by being there while the reaction is taking place, they make things go much faster. Here's an example – the breakdown of hydrogen peroxide, an unstable compound of hydrogen and oxygen. Left on its own, it will slowly break down into water and oxygen gas:

hydrogen peroxide → water + oxygen

Drop a spatula of powdered manganese dioxide into hydrogen peroxide and the powder starts to fizz rapidly as oxygen is given off. There is nothing particularly surprising in that, you might think – just another chemical reaction in progress.

Here's the surprise: if you filter the mixture after the reaction has finished you can get the same amount of manganese dioxide back. The oxygen has escaped and the water is left behind. The manganese dioxide has ended up just as it was. You can use it all over again if you want to.

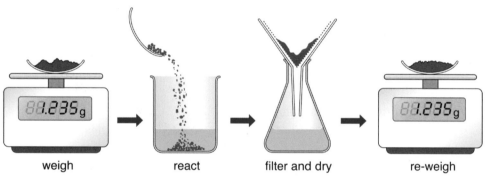

weigh   react   filter and dry   re-weigh

▲ Manganese dioxide speeds up the breakdown of hydrogen peroxide, but remains unchanged itself.

Weakening the bonds

The reaction that occurs is the simple breakdown of hydrogen peroxide to water and oxygen, which would have occurred slowly on its own. The manganese dioxide has simply speeded up this reaction without itself being altered. It has acted as a catalyst.

$$2H_2O_2 \xrightarrow{\text{MnO}_2 \text{ catalyst}} 2H_2O + O_2$$

Catalysts work by lowering the activation energy needed for the reaction. The process is complex but a simple model is that the existing bonds are weakened on the catalyst surface. This makes it easier for them to be broken during collisions so that new bonds can form. The overall result is that the reaction happens more easily.

> **Question**
>
> **a** Use the diagrams to explain how you could show that manganese dioxide is not taking part directly in this reaction.

> **Question**
>
> **b** Why do catalysts allow reactions to run at lower temperatures?

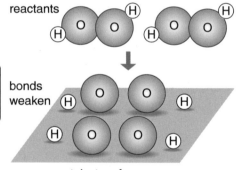

reactants

bonds weaken

catalyst surface

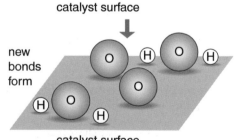

new bonds form

catalyst surface

products

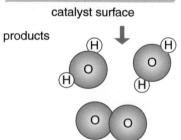

▲ A catalyst works by weakening bonds, which lowers the activation energy.

Surface area matters

This kind of reaction can only happen on the surface of the catalyst. So the catalyst has to be in the smallest pieces possible. Breaking it up into a fine powder may not help – the particles may just clump together. The answer is to attach small particles to tiny wire grids just millimetres apart. This allows the reactants to get to as large a surface as possible.

Nanotechnology (see section 1.8 in this unit) could take this approach to a new level. Create particles of catalyst just a thousandth of a millimetre in size, and their surface area would go up a thousand-fold. This would make them a thousand times as effective.

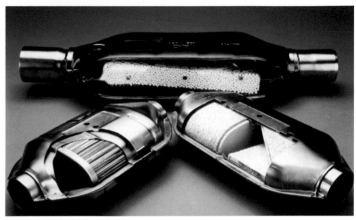
▲ The catalyst in a car's catalytic converter is held in a 'millimetre-sized' grid. Nanocatalysts could one day make cars completely pollution-free.

Question

c How could 'nanocatalysts' help revolutionise the way we tackle pollution?

The catalyst for the job

In the simple breakdown of hydrogen peroxide, iron oxide or copper oxide will work too, but manganese dioxide gives the fastest reaction. It is important to find the right catalyst for the particular reaction you want to speed up.

▲ Making chocolate like this needs a nickel catalyst.

Question

d How could you compare the effectiveness of different transition metal oxides as catalysts for the hydrogen peroxide reaction?

Transition metals or their oxides are often used as catalysts.

- Iron is used in the production of ammonia.
- Nickel is used to turn oils into fats for margarine or chocolate.
- Platinum is used in the production of nitric acid.
- Vanadium oxide is used in the production of sulfuric acid.

Questions

e In a particular industrial process, using a catalyst means that the reaction can run at 100 °C instead of 300 °C. How does this (i) save the company money (ii) help the environment?

f Platinum is very expensive. Why might this not matter so much for a catalyst?

Key points

- Catalysts speed up reactions without being used up themselves.
- Catalysts work by lowering the activation energy needed for the reaction.
- Transition metals or their oxides often make good catalysts.

Key to profitability

We rely on the chemical industry to provide the materials needed for our technology. It is vital to our way of life. The chemical industry, like any other business, can only exist if it makes a profit. This profitability depends on its ability to run the chemical reactions it uses as quickly, efficiently and cheaply as possible. By making reactions run faster at lower temperatures, catalysts dramatically reduce energy costs and so save money and increase profitability.

The transition metals and their oxides often make good catalysts. Different catalysts promote different reactions. Much work has gone into finding the 'best' catalyst for each process. Catalyst particle size is also important and nanoparticle catalysts are now beginning to revolutionise some chemical reactions.

> **Question**
>
> **a** Why do you think the chemical industry spends a lot of money on research and development of catalysts?

Precious and poisoned

Catalysts are useful but can have their problems. Some catalysts can be very expensive. Platinum is used as a catalyst for many reactions – but platinum is more expensive than gold. This should not be a problem as catalysts are in theory 'everlasting'. But sometimes catalysts are poisoned by impurities during the reaction. This happens when impurities attach themselves and stop the catalyst from working. Platinum, for example, was once used as a catalyst in the manufacture of sulfuric acid. Arsenic impurities quickly bonded to the platinum, and the precious metal had to be frequently replaced. Research led to the development of a new catalyst, vanadium oxide. This was not poisoned so easily and was cheaper too.

Gentle heating

Methanol catches fire if you heat it up and burns fiercely. A platinum catalyst can make it a safe, controlled heat source. Cordless hair tongs have a methanol fuel cartridge in the handle. When you press the button, the methanol mixes with air in a combustion chamber over a platinum catalyst on a fine mesh. The methanol reacts, giving off heat energy and warming up the tongs.

▲ Catalytic styling.

> **Question**
>
> **b** What other uses can you think of for this safe and easy source of 'instant heat' without the need for electricity or the risk of fire?

When and where we want it

Today we generate most of our electricity by burning fossils fuels. In the future we will get more and more of our electricity directly from hydrogen **fuel cells**. Fuel cells create electrical energy directly from hydrogen and oxygen with the help of a platinum catalyst, producing water as a waste product.

The basic reaction is: hydrogen + oxygen → water

But in the fuel cell each hydrogen atom splits up into a hydrogen ion (a proton) and an electron on the platinum catalyst. The proton passes across the membrane, but the electron goes up through the circuit. They meet up again on the other side, on another platinum catalyst. Here they react with oxygen to make water.

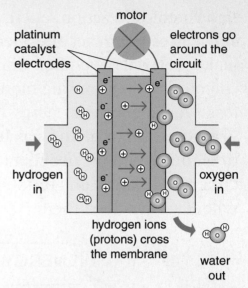

▲ Inside a fuel cell.

Questions

c Why are fuel cells better for the environment than traditional power stations?

d This sounds like the perfect, pollution-free energy source for the future. Hydrogen is often made from natural gas. Alternatively it can be made by the electrolysis of water – but how is the electricity made? If it comes from oil, gas or coal, you are still no better off. What sources of electricity could be used to make hydrogen that would not cause pollution? Is hydrogen the key to our future or not? Find out more and hold a class discussion.

Nature's catalysts

Plants use a catalyst to make photosynthesis work – it's called chlorophyll. In fact all living things, including you, rely on biological catalysts to work at all. We call them **enzymes** and they control all the chemical reactions in your body.

The chemical industry is also harnessing the power of enzymes to run some of its chemical reactions. It's an enzyme formed by yeast that helps turn sugar to alcohol in brewing, but other enzymes are used to turn corn starch to sugar syrup and milk to yogurt, for example.

Industrial food processing relies heavily on enzymes, but they are being put to other uses as well. Biological washing powders contain enzymes which break down grease and so help to clean clothes.

▲ Astronauts have been using fuel cells for years on the space shuttle for both electricity and drinking water.

Question

e Find out more about the use of enzymes in industry and evaluate their usefulness. What are their advantages and disadvantages compared to inorganic catalysts?

Key points

● Catalysts help industrial processes run faster and at lower temperatures.
● Catalysts help industry to reduce its energy costs.

In a chemical reaction, energy is transferred between the substances that react and their surroundings. When we make materials, we often draw on the energy stored in chemicals. They are like money in the bank. Some, like fossil fuels, got their energy from sunlight millions of years ago. Oil, coal and gas have been there for a long time, but when we use them up, they are gone forever. Others, like plant oils or foodstuffs, we can make afresh whenever we need them.

Balancing the books

We have to make withdrawals from the energy bank. We need petrol for our cars, paraffin for our aeroplanes and gas for our heating. We need food to power ourselves. We also need to make electricity to power our technological society. For example, we may need to supply energy to make the materials we need.

In this section we will be looking at the 'bank balance' of energy for the reactions we use to make materials.

Energy issues underlie all chemistry. They help to explain:

- why some chemical reactions happen 'on their own'
- why some chemical reactions have to the 'pushed'
- why temperature (and pressure) can affect the yield of reversible reactions.

Energy costs money. You need to get your 'energy sums' right if you want your business to make a profit. Even then, as long as we are reliant on fossil fuels, business is at the mercy of the global energy market. If the price of oil goes up, so does the cost of making things. How can chemists help to make their reactions more energy efficient? And can we find new energy sources?

Latest News!

Business hit by oil price hike!

Chemical stocks crashed yesterday as oil prices hit record highs ...

oil prices

stockmarket

▲ A better understanding of the way energy is used or released in chemical reactions has led to an enormous advance in battery technology. Just look at how mobile phones have changed over the last 30 years.

Energy out of control

Health and safety issues are important too. Some fuels are unstable and have to be handled with care. Coal is easy to store, but you have to be very careful with petrol, as the explosion and fire at the Buncefield fuel depot in 2005 shows (page 118). The study of energy in chemical reactions is very important if we are to deal with them safely.

Highly flammable
These substances easily catch fire.

Explosive
These substances cause an explosion.

▲ Look out for these hazard symbols.

Think about what you will find out in this section

What part does energy play in chemical reactions?

How do we keep track of the energy as reactions take place?

Why do some reactions give out energy while others absorb it?

How do energy changes affect the yield of reactions?

How are these ideas used in industry?

Energy on the loose

It is not always easy to tell when a chemical reaction has occurred. For example, if you mix cold, dilute hydrochloric acid and sodium hydroxide solutions, you will not see any obvious change. Two colourless liquids just mix to form another colourless liquid. Take hold of the tube and you will feel that something has happened. The tube has been warmed by the energy given out as the reaction takes place.

Chemical changes are often accompanied by changes in temperature, as energy is transferred to or from the surroundings. In this case, heat energy has been given out during **neutralisation**. Chemical reactions that give out heat energy like this are called **exothermic** reactions.

acid + alkali → salt + water + energy

e.g. HCl + NaOH → NaCl + H_2O + energy

Heat release runs in families

Many chemical reactions are exothermic like this – particularly reactions that start reacting as soon as the reactants are mixed. For example, metals reacting with acids:

metal + acid → salt + hydrogen + energy

e.g. Mg + H_2SO_4 → $MgSO_4$ + H_2 + energy

Displacement reactions are also exothermic. An example is when iron replaces the copper in copper sulfate solution:

iron + copper sulfate → copper + iron sulfate + energy

Fe + $CuSO_4$ → Cu + $FeSO_4$ + energy

◄ Iron displaces copper.

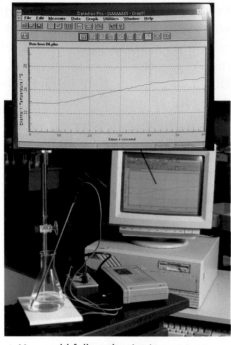

▲ You could follow the rise in temperature during neutralisation with a temperature sensor connected to a computer.

Question

a How else could you tell a reaction had occurred?

▲ Calcium and magnesium both fizz in acid. The test tubes will get hot.

Question

b Name two other metals that would react with acid.

Oxidation reactions, such as calcium metal turnings tarnishing in air, also give out energy:

$$\text{calcium} + \text{oxygen} \rightarrow \text{calcium oxide} + \boxed{\text{energy}}$$

$$2Ca + O_2 \rightarrow 2CaO + \boxed{\text{energy}}$$

Questions

c Describe a different exothermic reaction you have met.

d Do you think iron rusting is an exothermic reaction? Explain your answer.

▲ Lots of exothermic reactions, giving out heat and sound as well as light energy.

▲ An exothermic reaction that has got out of control.

A spark gets things going

Combustion reactions such as burning fuels are exothermic. For these reactions it's usually the energy we want, not the products. We burn methane to heat our homes and petrol to run our cars. Combustion reactions often produce light and sound as well as heat.

You have to light the fuel with a spark or match to start the reaction going. A kick-start of energy is needed to break some existing bonds before the reaction can occur. This is the activation energy needed for the reaction to begin. After that, the first reactions generate enough energy to keep the process going.

Questions

e Write equations for the combustion of coal (C), first in words and then using formulas.

f What problems might it cause if fuels did not need a kick-start of energy before they reacted?

g Petrol has a lower activation energy than paraffin. Why is it safe to store paraffin in an ordinary bottle, but petrol must be stored in a special safety container?

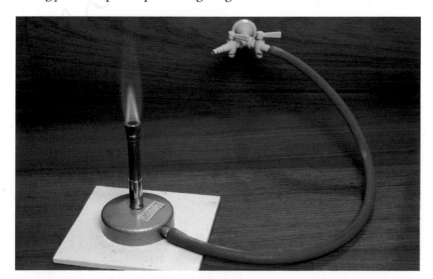

▲ Fuels need a kick-start to burn. Then their combustion reactions are highly exothermic.

Key points

- Fuels contain stored energy which is released when they burn.
- Fossil fuels contain energy collected from the sun and stored millions of years ago.
- Reactions that give out energy are called exothermic reactions.
- Many exothermic reactions such as burning need a 'kick-start' of energy from a spark or match.

The drink that cools itself

You can make a pleasant summer drink by adding a spoonful of baking soda to a glass of lemon juice. When you do this a chemical reaction occurs:

sodium hydrogencarbonate (baking soda) + citric acid (lemon juice) → sodium citrate + carbon dioxide + water

Question

a What would you see happening that would suggest that a chemical reaction was occurring?

If you were holding the glass as you stirred in the baking soda, you would also feel it get cooler. A temperature change is another indication that a chemical reaction is happening. In this case, the temperature goes down, not up.

▲ No ice is needed for this summer drink.

Supplying the energy to make things go

Reactions that take energy in from their surroundings are called **endothermic** reactions. It is fairly rare for endothermic reactions to 'go on their own' like this. You usually need to pump energy into them to make them work. Examples include materials that decompose when heated. We call these reactions thermal decomposition. Green copper carbonate ($CuCO_3$) will not break down on its own. Once heated, though, it quickly decomposes into black copper oxide (CuO) and carbon dioxide. The heat energy you put in breaks the compound apart.

Question

b Write word and balanced chemical equations for the decomposition of copper carbonate.

▲ Thermal decomposition reactions are endothermic.

One of the most important endothermic reactions of all is **photosynthesis**, in which plants use the energy from sunlight to build complex chemicals such as glucose from carbon dioxide and water:

carbon dioxide + water + **light energy** $\xrightarrow{\text{photosynthesis}}$ glucose + oxygen

▲ Sunlight energy powers this glucose factory.

The current way to purify metals

Electrolysis is an endothermic process. Electrical energy is used to split compounds into the ions that make them up. You can use electrolysis to extract aluminium from molten aluminium oxide, for example. It can also be used to make hydrogen and oxygen from water:

water + **electrical energy** → hydrogen + oxygen

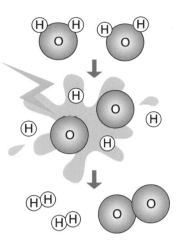

▲ Electrolysis is an endothermic process.

Questions

c The hydrogen and oxygen ions are held together by strong bonds in water. What is the electrical energy doing in this reaction?

d Balance the chemical equation for this reaction: $H_2O \rightarrow 2H_2 + O_2$

The chemical energy bank

Wind farms are a wonderful, pollution-free way of getting energy. The wind drives a turbine that generates electricity. There's one big catch. The wind doesn't blow all the time. Then, at other times, it blows so hard you make more electricity than you need. What can you do? Make a deposit in the energy bank. To do this, we have to find a way of storing the extra energy in chemical form.

▲ What's the big problem with wind farms?

Unused electricity on windy days can be used to electrolyse water and make hydrogen. This can then be stored until needed. When the wind stops blowing you can make an energy withdrawal. You can feed hydrogen into a fuel cell (see section 3.5 in this unit) to generate electricity. On a larger scale, this idea could be used to produce the hydrogen for the next generation of cars, lorries and buses. If hydrogen was produced this way, transport would become completely pollution-free.

Key points

- Many reactions require a continuous supply of energy to make them happen. They are called endothermic reactions.
- Photosynthesis is an endothermic process.
- Electrolysis, splitting a compound apart using electricity, is also endothermic.

Question

e Write a leaflet or design a poster to explain how hydrogen could become the totally pollution-free fuel of the future, from production to use.

More about reversible reactions

In section 2.6 of this unit you saw that blue copper sulfate has water molecules chemically bound up in its crystals. If you heat it, the water is driven out, forming white anhydrous copper sulfate.

If you add water to this white powder, it turns blue again. The water molecules are re-forming the chemical bonds and new, blue crystals form. If you did this in a test tube, you would feel the tube get hot. The energy is being given back as the bonds re-form.

▲ You need to put energy in to make white anhydrous copper sulfate.

▲ You get the same amount of energy back when you add water.

Question

a Is the reaction above exothermic or endothermic?

Question

b Is the reaction on the left exothermic or endothermic?

It all balances out

This reaction is reversible. The forward reaction, when bonds are broken, is endothermic. The back reaction, when bonds re-form, is exothermic.

forward reaction = endothermic

hydrated copper sulfate $\rightleftharpoons$ anhydrous copper sulfate + water

back reaction = exothermic

This is true of all reversible reactions. If the reaction is exothermic one way, it will be endothermic the other. And the same amount of energy will be gained or lost in each direction.

More about equilibrium

In a closed system, where no reactants or products are added or removed, reversible reactions 'stick' at a particular yield. That doesn't mean that the reaction has stopped – it just means that the forward and backward reactions are going at the same rate. For example, one molecule of pale yellow dinitrogen tetroxide (N_2O_4) breaks up to form two dark brown nitrogen dioxide molecules (NO_2) in a reversible reaction. At room temperature the reaction sticks just before it reaches the half-way point, giving the mixture a light brown colour overall.

Questions

c The other example of a reversible reaction given in section 2.6 was the action of heat on ammonium chloride, which breaks up (reversibly) into ammonia and hydrogen chloride. Describe this reaction in terms of the energy changes involved.

d Calcium carbonate forming calcium oxide and carbon dioxide is a reversible reaction.
$CaCO_3 \rightleftharpoons CaO + CO_2$
Which direction is endothermic and which is exothermic?

dinitrogen tetroxide $\rightleftharpoons$ nitrogen dioxide

$$N_2O_4 \rightleftharpoons 2NO_2$$

pale yellow **O O N N O O** $\rightleftharpoons$ **N O O N O O** dark brown

At equilibrium, dinitrogen tetroxide molecules are constantly breaking apart to form nitrogen dioxide molecules. At the same time other nitrogen dioxide molecules are colliding to form nitrogen dioxide molecules at the same rate.

Tipping the balance

The forward reaction as shown is endothermic, as energy needs to be put in to split up the larger molecule. The back reaction must therefore be exothermic. You get as much energy back as you had to put in to split up the larger molecule in the first place. This can be used as one way to control the equilibrium position of the reaction.

If you heat up the mixture, you will be driving the endothermic forward reaction, but working against the exothermic back reaction. This changes the amount of product made before equilibrium is reached. The more heat energy you put in, the more the reaction will be driven to the right. You will get more nitrogen dioxide and the mixture will get darker. If you cool the mixture down, you will be discouraging the forward reaction. The back reaction will be encouraged, as the heat energy it gives out will be removed by the cooling. You can see one effect of the change – the mixture will get lighter in colour.

Questions

e In a closed system at equilibrium, the forward reaction was converting 0.1 g of dinitrogen tetroxide to nitrogen dioxide every second. What was the rate of the back reaction?

f Which reaction (forward or back as shown) is a thermal decomposition reaction?

Question

g Explain how the diagram supports the idea that changing the temperature will change the balance point in this reaction.

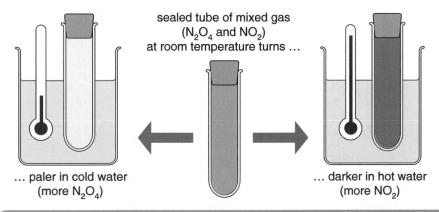

sealed tube of mixed gas (N$_2$O$_4$ and NO$_2$) at room temperature turns …

… paler in cold water (more N$_2$O$_4$)

… darker in hot water (more NO$_2$)

Key points

- In reversible reactions, one direction is exothermic, the other is endothermic.
- Changing the temperature changes the position of equilibrium in a closed-system reversible reaction.

Temperature and yield

You have met the Haber process before in section 2.7 of this unit. The reaction between nitrogen and hydrogen can give a good yield of ammonia at room temperature, but the reaction is very slow. One obvious way to speed this up is to heat it up. In general you double the rate of a reaction if you raise the temperature by 10–15 °C. You would need to raise the temperature to 1000 °C or more to make this reaction run fast enough.

Raising the temperature increases energy costs, which is never good for business or the environment. In this reaction there was an even bigger problem. The reaction is reversible. The forward reaction is exothermic and the back reaction is endothermic:

forward exothermic

$$N_2 + 3H_2 \rightleftharpoons 2NH_3$$

back endothermic

In many ways, reversible reactions are like awkward characters. Whatever you do to them, they try to do the opposite to counteract it. When you heat up the reactants, you put energy in, which the reaction tries to absorb. Then the endothermic back reaction runs faster so the yield of ammonia gets worse. Raising the temperature always drives the reaction in the endothermic direction.

This is a real problem for the reaction. You can run it cold and get a good yield, but the reaction then runs far too slowly to be of any use. Even with a good catalyst, the temperature has to be so high that the yield is always poor.

The pressure effect

The volume of a gas depends on how many particles there are (see section 2.4 of this unit). In this reaction, one nitrogen molecule plus three hydrogen molecules react to give just two molecules of ammonia. At any given temperature, if this reaction went to completion, its volume would halve.

1 mole nitrogen + 3 moles hydrogen → 2 moles ammonia

1 volume   3 volumes   2 volumes

In this case, if you increase the pressure, the reacting mixture responds by getting smaller. It does this by shifting the equilibrium to the right – you get a better yield of ammonia.

> **Question**
>
> *a* Why is it always better to run industrial processes at as low a temperature as possible
> *(i)* for business
> *(ii)* for the environment
> *(iii)* for sustainability?

> **Question**
>
> *b* If 1 dm³ of nitrogen and 3 dm³ of hydrogen reacted completely, what volume of ammonia would you get (at the same temperature and pressure)?

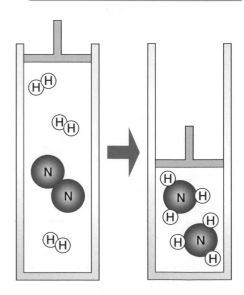

▲ Half the number of particles – half the volume.

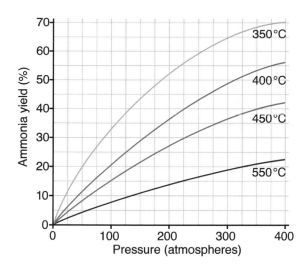

▲ How ammonia yield varies with temperature and pressure.

The graph shows how ammonia yield varies with temperature and pressure in the Haber process.

Industrial efficiency

A typical modern Haber plant runs at 450°C (which means the reaction runs fast enough, using an iron catalyst) and 200 atmospheres pressure. This only gives just under a 30% yield within its closed, pressurised reaction vessel. But remember that Haber's great breakthrough was to 'open' the system and remove the ammonia by cooling and liquefying it (see section 2.7 of this unit). The unreacted gases are then 'topped up' and recycled, until eventually all the reactants are turned into ammonia.

The Haber process runs continuously, with the reactant gases piped in and the liquid ammonia piped out. It has been fine-tuned over the years to become a very efficient process, reducing the price of ammonia and so fertilisers, while still making a good profit for the companies.

Questions

c What yield of ammonia would you get at 250 atmospheres pressure at (i) 350°C (ii) 450°C?
d What yield would you get at 400°C at (i) 200 atmospheres pressure (ii) 400 atmospheres pressure?
e What yield would you predict at 400°C and 600 atmospheres?
f Suggest reasons why this reaction is not run at such a high pressure.
g Why is a high temperature used, even though it reduces the yield?

Questions

h Roughly how many times would the gases have to cycle to get complete conversion?
i What costs, other than energy costs, are kept low by this efficient, continuous process?
j The actual yield (over many cycles) is effectively 100%. But what is the atom economy of this reaction? (See section 2.8 of this unit.)

Key points

- Lowering the temperature of the Haber process increases the yield, but slows the reaction.
- Increasing the pressure increases the yield but can be very expensive.
- Over the years, this process has been 'fine-tuned' to make the best of these apparently conflicting factors.

Waste not, want not

Good for business ...

In the early days of the chemical industry, energy was relatively cheap and there was no thought that our fossil fuels might run out, or that we might be causing major damage to the environment. Industrial cities were blackened by smoke from the factories and hundreds of thousands of people died prematurely because of lung disease. Acid rain fell, damaging buildings and killing trees and the fish in lakes. Toxic waste poured into many rivers, making them foul and lifeless.

... better for the planet?

Today the outlook is different. We know the benefits of the chemical industry to our everyday lives, but we also know more about the costs. We know that:

- fossil fuels will run out, possibly in your lifetime
- burning fossil fuels releases the greenhouse gas carbon dioxide
- 'wasted' heat energy causes thermal pollution that can affect local ecosystems such as those in lakes.

How can we change the way industry works? How can we make it sustainable for future generations? One way is through education – we are all now more aware of our responsibilities towards the Earth. Many companies have latched on to this idea and are trying to change their image to attract more people to their brand.

A second way is through legislation. We have laws that force companies to behave responsibly towards the environment, and penalise them if they cause pollution. That is more difficult to get in place globally, however.

A third way works hand in hand with business's efforts to make money. As fossil fuel supplies dwindle, their price goes up. Anything industry can do to reduce its energy costs will save money and improve the chances of making a profit. Whatever the motive, if industry uses less energy we will burn less fossil fuel and the whole planet will benefit.

Question

a How has the environment of many towns and cities improved since this photo was taken in 1946?

Question

b BP (British Petroleum) is an oil company. What image is this advert trying to give to it?

Question

c Discuss the relative merits of the three approaches to changing the chemical industry for the better. Which is best? How should they be balanced? Write your own 'manifesto for the future of the chemical industry'.

Chemists conserve energy

Sulfuric acid is a very important chemical that is used to make paints, fertilisers, plastics and detergents. A key part of its production involves converting sulfur dioxide to sulfur trioxide. The reaction is reversible, so you have to get the conditions right. In this reaction, the forward reaction is exothermic, the back reaction endothermic:

sulfur dioxide + oxygen $\rightleftharpoons$ sulfur trioxide

$$2SO_2 \quad + \quad O_2 \rightleftharpoons \quad 2SO_3$$

Because of this, the best yields are obtained at low temperatures. But the reaction is far too slow then to be of any use industrially.

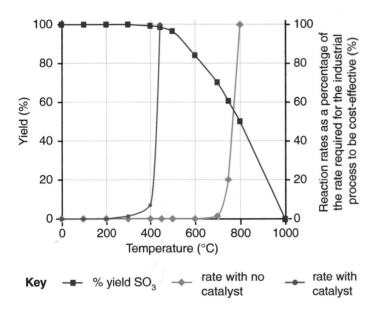

Key — ■ % yield SO₃ ◆ rate with no catalyst ● rate with catalyst

▲ Yield and reaction rate graph for the production of sulfur trioxide.

In some industrial reactions increased pressure could reduce the temperature needed for the reaction, but very high pressure vessels are very expensive to build (and run) safely. The current potential fuel savings do not warrant the extra capital expense. This view might change in the future as fossil fuel sources become scarcer and the price inevitably rises.

Questions

Use the graph to help you answer these questions.

d What yield would this reaction give at 600 °C (red line only)?

e Without a catalyst (green line), what temperature would be needed to make this reaction fast enough (100%)?

f What yield (red line) would you get at this temperature?

g With a vanadium oxide catalyst (blue line) the reaction can run successfully at 450 °C. What yield (red line) will this give?

h Energy costs go up dramatically as the temperature increases. Suggest two reasons why companies making sulfuric acid invested a lot of money in finding a good catalyst.

i With a yield that high, you don't need to worry about the pressure. But would high pressure improve the yield or make it worse? (Hint: you will need to look at the number of particles.)

j This reaction could be run at high pressure and high temperature without a catalyst. Why do you think a catalyst is used instead of high pressure?

Question

k Write a letter to your boss in the chemical company, outlining the reasons why money should be spent on the research and development of new catalysts for industrial processes.

Key points

● A detailed understanding of the science behind chemical reactions can help to reduce energy costs for the chemical industry.

● This is good for the industry, but it also helps to make it sustainable and so is good for the planet.

Mention salt and you probably think of sodium chloride – in sea water, on your chips, in your dishwasher – or even on the road in winter. Sodium chloride is one member of a very important family of compounds. Chemists call them all **salts**. Salts are all ionic compounds. They usually have positive metal ions and negative non-metal ions.

◀ Sea water contains 2.8% of sodium chloride – plus 1% of other salts.

What do all salts have in common?

In solid form, the ions are arranged on a giant ionic lattice. This means that they:

- are hard but brittle
- have high melting points
- do not conduct electricity when solid
- often dissolve in water – the solutions conduct electricity.

Because the ions carry electrical charge, and are relatively free in solution, it is fairly easy to separate the ions and use them to make different materials. Such processes play an important part in the way the cells of our bodies work. We can also use electrical forces to separate the ions from salts and similar substances in order to obtain materials such as metals.

cobalt chloride

copper sulphate

chromium chloride

nickel nitrate

iron chloride

◀ These are all salts.

Mined from the ground

Most ores are ionic compounds. They consist of giant ionic lattices that include ions of metals and oxygen or sulfur. They differ from salts in some respects – for example, they may not dissolve in water.

Less reactive metals can be obtained from their ores by carbon reduction. This does not work for more reactive metals such as aluminium. Aluminium has to be ripped from bauxite, its oxide ore, by electrolysis. This only works if the bauxite is heated until it melts. Molten ionic materials conduct electricity.

zinc blende
(ZnS)

bauxite
(Al_2O_3)

▲ Metal ores are ionic too.

Life-sustaining ions

Your body and your very life depend on ions in solution. Sodium ions control the water balance in your cells. Too much salt is bad for you – but so is too little. Your dog or cat licks your skin to get the salt they need from your sweat.

Potassium salts are important in the way your nerves work. They are also important for your brain – you need a good supply of potassium ions to think properly. Bananas contain potassium ions, so they make a great snack when you're revising.

Calcium salts are important for your bones and teeth. Milk and cheese will give you plenty of calcium. Acids in liquid form are also ionic. The greater the concentration of hydrogen ions, the stronger the acid. You have strong hydrochloric acid in your stomach to help digest your food.

Think about what you will find out in this section

We can use the charge on ions as a way of controlling them.	How to make different salts from acids and alkalis.
We can use electrical energy to extract ions from salts.	The world of materials made by electrolysis.
What you can get from sea water.	You will also have a chance to practise balancing equations and calculating reacting masses.

Liquefy and liberate

Solid ionic materials have a giant ionic structure of positive and negative ions. Strong electrostatic forces hold the ions in place, so ionic solids do not conduct electricity. If the solid melts or dissolves in water, the ions break free. Molten or dissolved ionic materials conduct electricity as the ions themselves move.

Elements from molten salt

When you melt sodium chloride, the positive sodium ions (Na^+) and the negative chloride ions (Cl^-) break free from the ionic structure. Put electrodes into molten sodium chloride, and pass a current through the liquid. Electrons build up on the negative electrode, attracting the positive sodium ions. When each ion reaches the electrode, an electron combines with it, cancelling out the charge. The sodium ion turns back into a sodium atom. Metallic sodium is deposited on the negative electrode.

Meanwhile, the negative chloride ions are attracted to the positive electrode. When each ion reaches the electrode, its extra electron is ripped from it. The chloride ion turns back into a chlorine atom. Pairs of atoms then join up and chlorine gas bubbles out on the positive electrode. This process is called **electrolysis**.

A game of two halves

Electrolysis reactions can be written down as chemical equations. It is often easier to write these as **half equations**, taking the reaction at each electrode in turn. For the electrolysis of molten salt:

at the negative electrode

1 positive sodium ion + 1 electron → 1 sodium atom

$$Na^+ \quad + \quad e^- \quad \rightarrow \quad Na$$

This is a form of **reduction**.

at the positive electrode

2 negative chloride ions → 1 chlorine molecule + 2 electrons

$$2Cl^- \quad \rightarrow \quad Cl_2 \quad + \quad 2e^-$$

This is a form of **oxidation**.

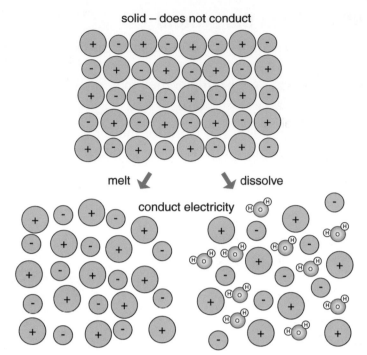

▲ Ionic materials only conduct electricity when molten or in solution.

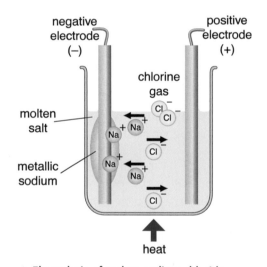

▲ Electrolysis of molten sodium chloride.

Questions

a Why are the negative ions attracted to the positive electrode?

b If you electrolysed molten potassium chloride, what would you get at the negative electrode?

c What kind of chemical bond holds the two chlorine atoms together in a chlorine molecule?

d Is electrolysis an exothermic or endothermic process?

Question

e Why can't sodium be made by carbon reduction like iron or copper?

▲ Sodium metal is produced commercially by electrolysis.

OilRig: adding and removing electrons

In the simplest form, oxidation just means adding oxygen, and reduction means taking oxygen away. But these terms are used widely in chemistry. A better definition uses the ideas shown for electrolysis. **O**xidation **is** the **l**oss of electrons, **r**eduction **is** the **g**ain of electrons. Just remember **O i l R i g**!

Electrolysis in action

Electrolysis is used to get reactive metals such as aluminium from their ores. Molten aluminium oxide ore (bauxite) is ripped apart using large amounts of electricity. Aluminium electrolysis plants have to have their own power stations.

Questions

f Aluminium is in Group 3 of the periodic table and its atomic number is 13. How many electrons does it have in its outer electron energy level?
g What will the charge on an aluminium ion be if it loses all of these electrons?
h If molten aluminium is electrolysed, which electrode will the aluminium ion be drawn towards?
i Write a half equation for the reaction of the aluminium ion at this electrode.
j What is the % of aluminium by mass in aluminium oxide (Al_2O_3)? (Al = 27, O =16)
k How much aluminium could you get by electrolysis from 1 tonne of bauxite?

▲ An aluminium refinery with its own hydroelectric power station.

Key points

- Ionic compounds can be separated by electrolysis once the ions have been freed up by melting or solution.
- During electrolysis positive ions go to the negative electrode and gain electrons, negative ions go to the positive electrode and lose electrons.
- You can define oxidation and reduction in these terms; remember OilRig!

Useful chemicals from sea salt

Salt occurs naturally in solution in the sea – **brine** – and in large beds of rock underground – **rock salt**. When salt is dissolved in water, it splits up into its ions (Na^+ and Cl^-), which are free to move about. Because of this, brine conducts electricity. This idea is used in industry to split the brine apart by the process of electrolysis.

But the picture is complicated as the water also ionises to give H^+ and OH^- ions. If you electrolyse a solution with mixed ions, the less reactive ions are the ones that are changed. So in this case you get hydrogen not sodium. The result is three useful products.

▲ When seas like the Great Salt Lake of Nevada USA dry up, they leave beds of salt behind.

- Hydrogen gas forms at the negative electrode. This is used to make ammonia for fertilisers, and change oils into fats for margarine and chocolate. It can also be used as a fuel. The half equation is:
$$2H^+ + 2e^- \rightarrow H_2$$

- Chlorine gas forms at the positive electrode. Chlorine ions kill bacteria in swimming pools and drinking water, and are also used to make disinfectants, bleach and plastics such as PVC.

- That leaves sodium and hydroxide ions in solution. Sodium hydroxide is a strong alkali also know as caustic soda. It is used to make soap, paper and ceramics. It is also used to clean ovens.

Overall the reaction is:

$$\text{sodium chloride} + \text{water} \xrightarrow{\text{electrolysis}} \text{chlorine} + \text{hydrogen} + \text{sodium hydroxide}$$

$$2NaCl\,(aq) + 2H_2O\,(l) \longrightarrow Cl_2\,(g) + H_2\,(g) + 2NaOH\,(aq)$$

▲ Thick beds of salt in the rocks below Cheshire UK show that this happened in the past, too.

Symbols in brackets show the state of a reactant in chemical formulae. If it's in solution, (aq) is used. Otherwise, (l), (s) and (g) indicate liquid, solid and gas. They are called **state symbols**.

Questions

a Write the half equation for the formation of chlorine at the positive electrode.
b Are the hydrogen ions oxidised or reduced?
c Why isn't metallic sodium produced at the negative electrode? (Hint: think what happens if you put sodium in water.)

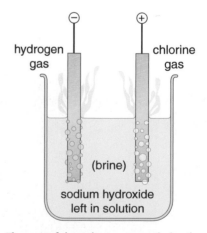
▲ Three useful products are made by the electrolysis of brine.

Shiny metal from deep blue liquid

Copper has positive ions in copper sulfate solution. These ions give copper sulfate its blue colour. In copper sulfate solution the ions are free to move. If you put electrodes into copper sulfate, the positive copper ions are attracted to the negative electrode. When they reach the electrode, the charges cancel out and the copper ions turn back to copper atoms. Metallic copper is deposited on the electrode.

> ### Questions
>
> **d** Why are the copper ions attracted to the negative electrode?
> **e** What would happen to the colour of the solution if all the copper ions turned back to metal atoms at the negative electrode?

Purifying copper

Copper is usually produced by reduction of its oxide ore in a furnace. The metal produced is not pure enough for commercial use. Electrolysis is then used to purify it further. This is cheaper than you might expect because the ions involved can be freed up in water, without the expense of high temperatures.

A block of impure copper is connected up as the positive electrode in a solution of copper sulfate. A strip of pure copper is used as the negative electrode. When the current is switched on, the positive copper ions are attracted to the negative electrode and deposited there.

Meanwhile, the copper atoms from the impure copper block become positive ions and go into solution to take their place. Eventually, the block disappears. The solution acts like a 'conveyor belt', moving the copper atoms across from the positive to the negative electrode. Eventually all of the copper ends up on the negative electrode – as very pure copper indeed.

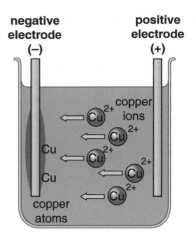

▲ Electrolysis of copper sulfate solution.

> ### Questions
>
> **f** Impure copper has a higher electrical resistance than pure copper. Suggest a use of copper that needs it to be very pure.
> **g** There are hydrogen ions in the solution. Why is copper formed, not hydrogen?

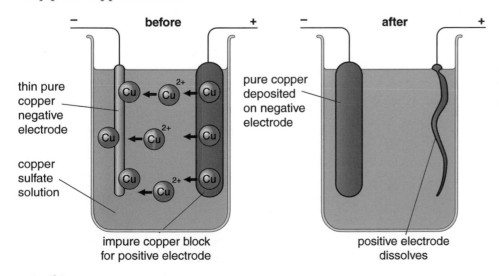

▲ Purifying copper.

> ### Key points
>
> - When sodium chloride solution is electrolysed, hydrogen and chlorine gases are produced at the electrodes. Sodium ions and hydroxide ions are left in the solution.
> - Copper made in a furnace from copper ore can be further purified by electrolysis.
> - State symbols in an equation show solid (s), liquid (l), gas (g) or solution in water (aq).

Combinations in everyday life

You know about the everyday uses of common salt, sodium chloride. But many other salts have important uses, for example:

- potassium nitrate is used in gunpowder
- calcium sulfate is used in Plaster of Paris
- copper sulfate is sprayed on grapevines to prevent disease.

There are many different ways to make 'the salt for the job'. This section and the next look at some of those ways.

Making salts straight from the metal

Salts are ionic compounds that have a positive 'metal' part, and a negative part from an acid. One way to make a salt is to place a metal in the appropriate acid:

metal + acid → salt + hydrogen

You can make magnesium sulfate, for example, by reacting magnesium ribbon with sulfuric acid:

magnesium + sulfuric acid → magnesium sulfate + hydrogen

$$Mg + H_2SO_4 → MgSO_4 + H_2$$

In this reaction, the metal is displacing hydrogen from the acid. It only works for metals that are more reactive than hydrogen.

If you held this mixture in a test tube you would see the hydrogen bubbling out. You would also feel the tube getting hot. Most acid reactions that make salts are exothermic.

▲ Copper sulfate solution in use in the vineyard.

very reactive

potassium
sodium
calcium
magnesium
aluminium
carbon
zinc
iron
tin
lead
hydrogen
copper
silver
gold
platinum

unreactive

▲ The reactivity series.

Questions

a Write a balanced equation for the production of zinc sulfate ($ZnSO_4$) from zinc and sulfuric acid.

b Could you make copper sulfate in this way? Explain your answer.

c Why would it not be a good idea to make sodium sulfate in this way?

d Which acid do you think potassium nitrate is made from?

Making salts by changing partners

Sodium iodide and lead nitrate both dissolve in water. But mix them together and a beautiful yellow solid appears. This **precipitate** of insoluble lead iodide can then be filtered off, washed and dried. The two salts have 'swapped partners' in what is called a **double decomposition** reaction.

sodium iodide + lead nitrate → lead iodide + sodium nitrate

(clear solutions)   (yellow precipitate)   (clear solution)

$2NaI\,(aq)$ $+ Pb(NO_3)_2\,(aq) →$ $PbI_2\,(s)$ $+$ $2NaNO_3\,(aq)$

This happens because the individual solutions contain free ions. When they mix, the lead (Pb^{2+}) and iodide (I^-) collide, stick together and fall out of solution as a solid mass.

Any insoluble salts can be made in this way, by mixing solutions of the appropriate soluble compounds.

▲ Precipitation of yellow lead iodide from aqueous solution.

Questions

e   *How many iodide ions combine with each lead ion?*
f   *What do the symbols (aq) and (s) mean in the equation?*
g   *Sodium chloride and silver nitrate are soluble, but silver chloride is not. How could you make a sample of silver chloride?*

The bad stuff just drops out

Double decomposition is used to clean up polluted water. Effluent from factories often contains dissolved ions of transition metals such as copper, chromium, cadmium or mercury. Sodium carbonate is soluble in water, but transition metal carbonates are insoluble.

If the effluent is treated with sodium carbonate, the insoluble metal carbonate will precipitate out and can be filtered off and removed. So for copper:

copper ions + carbonate ions → solid copper carbonate

$Cu^{2+}\,(aq)$ $+$ $CO_3^{2-}\,(aq)$ $→$ $CuCO_3\,(s)$

▲ Metal ions in factory effluent can be highly toxic and damage the environment.

Question

h   *Iron carbonate is insoluble in water. How can iron ions be removed from drinking water?*

Key points

● Metal salts have many uses and can be made in various ways.
● Some salts can be made by reacting the metal with the appropriate acid.
● Insoluble salts can sometimes be made by getting soluble salts to 'change partners' in a double decomposition reaction.

Other ways to make salts

Acids and alkalis: dangerously reactive

All acids contain hydrogen. In aqueous solution the hydrogen is present in the form of free hydrogen ions (H^+). The greater the concentration of hydrogen ions, the stronger the acid.

Alkalis are soluble metal hydroxides. In aqueous solution the hydroxide ions (OH^-) are freely available. The greater the concentration of hydroxide ions you have, the stronger the alkali. Strong acids and strong alkalis are both very reactive – they are dangerous. That's why we have to be very careful storing and using them.

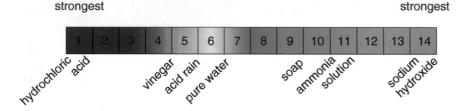

◄ Colour changes with universal indicator: 1 is the strongest acid, 14 the strongest alkali. Water has a pH of 7. This is neither acidic nor alkaline – it is neutral.

The strength of an acid or alkali is measured against the **pH scale**. This scale runs from 1 to 14. The pH of a solution may be measured using **universal indicator**. This changes through many different colours from pH 1 to 14.

Bringing opposites together

Acids and alkalis are chemical opposites. Mixed together in the right proportions they can cancel one another out. This process is called neutralisation. For example:

sodium hydroxide + hydrochloric acid → sodium chloride + water

$$NaOH\,(aq) \quad + \quad HCl\,(aq) \quad \rightarrow \quad NaCl\,(aq) \quad + H_2O\,(l)$$

> **Question**
>
> **a** What colour change would you see if you poured acid into alkali mixed with universal indicator until neutralisation occurred? What if you kept adding acid?

To get pure sodium chloride you need to react equal molar quantities of the acid and alkali. You get this if you mix equal volumes of the same concentration of the reacting substances. You could then get salt crystals by carefully boiling the water away. When there is not much liquid left, you may want to turn off the gas (to avoid spitting) and let the last of the water evaporate safely.

You can make different salts using different acids and alkalis. The name of the salt depends on the metal in the alkali and the acid used.

- Hydrochloric acid (HCl) gives chloride salts.
- Sulfuric acid (H_2SO_4) gives sulfate salts.
- Nitric acid (HNO_3) gives nitrate salts.

▲ Getting salt crystals from solution.

> **Question**
>
> **b** How could you make (i) sodium sulfate (ii) potassium nitrate?

Insoluble bases

Only water-soluble hydroxides make alkalis. Many metal hydroxides, like those of the transition metals, are insoluble and so are not alkaline. But they can still react with acids to form salts.

Alkalis are water-soluble examples of a larger group of chemicals that can neutralise acids. These are called **bases**. The neutralisation reaction is:

acid + base → salt + water

The key reaction at the heart of these neutralisations is always the same. The hydrogen ions and hydroxide (or oxide) ions join together to make water. For example:

$$H^+ (aq) + OH^- (aq) \rightarrow H_2O (l)$$

The metal oxide route

Metal oxides are bases too. Salts of transition metals can be made by reacting the metal oxide with the appropriate acid. So for copper sulfate:

copper oxide	+	sulfuric acid	→	copper sulfate	+	water
(black solid)		(colourless solution)		(blue solution)		
$CuO (s)$	+	$H_2SO_4 (aq)$	→	$CuSO_4 (aq)$	+	$H_2O (l)$

hydrogen from the acid

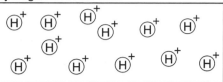

oxygen from the base

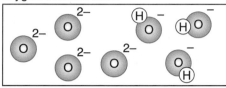

hydrogen and oxygen form water

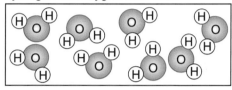

▲ Making water.

Question

c *Magnesium hydroxide is slightly soluble in water. Is it (i) a base (ii) an alkali?*

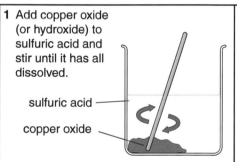

1 Add copper oxide (or hydroxide) to sulfuric acid and stir until it has all dissolved.

sulfuric acid

copper oxide

2 Add more copper oxide a little at a time until no more will react and dissolve.

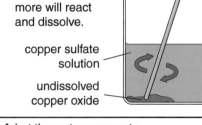

copper sulfate solution

undissolved copper oxide

3 Filter out the unused copper oxide. The clear filtrate will be copper sulfate solution.

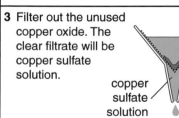

copper sulfate solution

4 Let the water evaporate away to give blue copper sulfate crystals.

copper sulfate crystals

Questions

d *What salt would you get if you added copper oxide to hydrochloric acid?*
e *Describe how you could make nickel sulfate ($NiSO_4$) from insoluble nickel oxide (NiO). Write word and balanced chemical equations for this reaction.*

Key points

- The strength of acids and alkalis is measured on the pH scale. Universal indicator changes colour with this scale.
- Acids contain free hydrogen ions (H^+). The higher the concentration, the stronger the acid.
- Chemicals that can neutralise acids are called bases.
- Soluble bases are called alkalis. The higher the concentration of hydroxide ions (OH^-), the stronger the alkali.
- Acid plus a base forms a salt plus water.

Ammonia and the fertiliser debate

The strange case of ammonia

Ammonia is a gas with a very strong smell. It is a simple compound of the non-metals nitrogen and hydrogen (NH_3). When it dissolves in water, it forms a compound called ammonium hydroxide:

ammonia + water → ammonium hydroxide

$$NH_3 (g) + H_2O (l) → NH_4OH (aq)$$

Ammonium hydroxide solution is a reasonably strong alkali. It is often used as a kitchen cleaner, as alkalis are good for clearing up grease. The 'ammonium' (NH_4^+) part contains no metal, and yet it is acting as if it was an alkali metal. For example, it can be used to neutralise acids, forming 'ammonium' salts with them.

By reacting nitric acid with ammonium hydroxide we can make ammonium nitrate (NH_4NO_3). Ammonium nitrate is used as a fertiliser as it contains a lot of nitrogen in a form that plants can use. Plants need nitrogen to grow well.

The fertiliser debate

People were worried that they could not grow enough crops as the global population approached 1½ billion a hundred years ago. Making 'artificial' fertilisers from the nitrogen in the air seemed like an ideal solution (see section 2.7 of this unit).

Today the population is four times as high. You might think that ammonium nitrate fertiliser would be seen as the answer to feeding the world. Some people have their doubts. Read the arguments, for and against, on the next page.

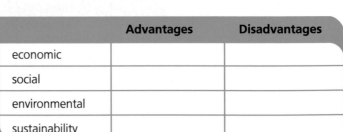
▲ Ammonium nitrate – the fertiliser that gives the most nitrogen per kilogram.

Questions

a How many nitrogen atoms are there in one 'particle' of ammonium nitrate?

b What is the formula mass of ammonium nitrate? ($H = 1$, $N = 14$, $O = 16$)

c What is the percentage of nitrogen in ammonium nitrate by mass?

d How many grams of nitrogen will you get for every kilogram of fertiliser?

e Ammonium hydroxide + nitric acid → ammonium nitrate + water. The ammonium ion (NH_4^+) acts just like Na^+. Write a balanced equation for this reaction.

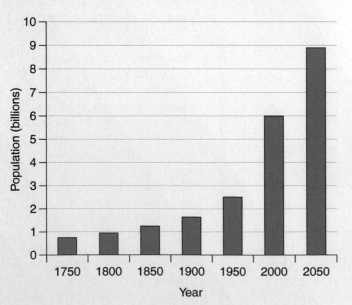

	Advantages	Disadvantages
economic		
social		
environmental		
sustainability		

Question

f Copy the table. Draw up a list of the advantages and disadvantages of using ammonia-based fertilisers, using the headings shown.

▲ World population, showing the expected increase by 2050.

Ammonium nitrate is cheap and easy to make on a large scale. We get the nitrogen 'free' from the air.

We can get it by electrolysing water.

Ammonium nitrate contains 35% nitrogen which plants can use directly. Farming yields rise dramatically when we use it. How could we feed the world without it?

Yes, but the population has more than doubled over the last 50 years. It could do the same in the next 50. Compost and manure are fine for small farms or gardens, but they are just not suitable for the large-scale 'industrial' farming that we need to feed the world.

The use of manure can contaminate food with bacteria. Artificial fertilisers are disease-free.

The solubility of ammonium nitrate helps the plants to get at it. You don't need to use that much to get the effect. Use less fertiliser, save money and avoid the pollution. It's just a matter of good farming practice.

Yes, but you get the hydrogen from natural gas – what happens when that runs out?

And how is the electricity made? It's not a problem if it's from a renewable source, but today it's mostly from burning fossil fuels. That's just not sustainable.

Farmers can get all the nutrients they need back into the soil by traditional, organic farming methods. Compost and manure have been used for millennia.

Large-scale farming like this has not been good for developing countries where the population is growing fastest. We need sustainable farming where local communities meet their own needs.

Ammonium nitrate fertiliser washes out of the soil when it rains. It gets into rivers and causes algae to grow. These 'algal blooms' can kill off all other life in rivers or lakes.

If only it was that easy! And it can get into drinking water too. High nitrate levels can be very bad for babies.

Key points

- Ammonium salt fertilisers allow us to grow enough crops to feed the world, but some people are concerned about the negative effects.
- For sustainable development we need to be sure that we are working in tune with nature and that we are not storing up problems for the future.

1 Copy and complete this table for sub-atomic particles. *(3 marks)*

Particle	Charge	Position
	+1 (positive)	nucleus
neutron	0 (neutral)	
electron		whizzing around in energy levels

2 Look at the simple periodic table:

H							He

Li	Be	B	C	N	O	F	Ne
Na	Mg	Al	Si	P	S	Cl	Ar
K	Ca						

a Which element has
  i four electrons in its second energy level? *(1 mark)*
  ii seven electrons in its third energy level? *(1 mark)*
  iii two electrons in its fourth energy level? *(1 mark)*

b For each compound, say whether it forms by 'sharing' or 'give and take' of electrons.
  i NO_2 **ii** K_2O **iii** NaF   **iv** CCl_4
  (4 marks)

c Name an element that has a full outer energy level. *(1 mark)*

3 a Draw the electronic structure of an oxygen (O) atom. *(2 marks)*

b How many electrons does it need to fill its outer energy level? *(1 mark)*

c What would the charge be on this ion when the outer energy level is full? *(1 mark)*

d Which noble gas has the same electron structure as this oxygen ion? *(1 mark)*

e Draw the electron structure of a magnesium (Mg) atom. *(2 marks)*

f What would the charge be on this ion if it lost its outer electrons? *(1 mark)*

g Which noble gas has the same electron structure as this magnesium ion? *(1 mark)*

h What type of compound would oxygen and magnesium form? *(1 mark)*

i Given your answers to questions **c** and **f**, what would you expect the formula of magnesium oxide to be? *(1 mark)*

4 Look at the table of properties for materials **A** and **B**.

Property	A	B
conduction of electricity	conducts when solid or liquid	does not conduct when solid; conducts when liquid
hardness/strength	may be hard, but can be easily bent or shaped	hard but brittle
melting point	high	high

For **A** and **B** in turn:

a Does it have a metallic or ionic compound structure? *(2 marks)*

b Draw a simple sketch of this structure. *(4 marks)*

c Given your answer to **a**, explain each of the three properties. *(6 marks)*

5 a Copy and complete this table showing some non-metals and their compounds. *(2 marks)*

Element	Symbol	Electron structure	Formula of compound with hydrogen
carbon	C	2,4	
nitrogen	N	2,5	NH_3
oxygen	O	2,6	H_2O
fluorine	F	2,7	

b What type of bonding joins the atoms together in all these compounds? *(1 mark)*

c Draw molecular models to show all four molecules. *(4 marks)*

d Draw 'dot and cross' diagrams for the outer electrons to show how each of these compounds forms. *(4 marks)*

6 The diagrams show the structure of two non-metal elements, carbon (diamond) and oxygen.

carbon (diamond)

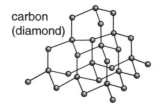

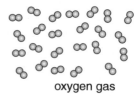

oxygen gas

a
  i What type of bonding holds the carbon atoms together in diamond's giant structure? *(1 mark)*

  ii Explain why diamond is very hard and has a very high melting point. *(2 marks)*

b
  i What type of bonding (force) holds the oxygen atoms together *within* the molecules? *(1 mark)*

  ii What type of force exists *between* the molecules? *(1 mark)*

  iii Explain why oxygen is a gas at room temperature. *(2 marks)*

7 State whether each substance described below is metallic, ionic, molecular covalent or giant covalent. Explain your answer in each case.

a **W** is a liquid that easily vaporises. It does not conduct electricity. *(2 marks)*

b **X** is a hard but brittle solid that does not conduct electricity. **X** dissolves in water and the solution does conduct electricity. *(2 marks)*

c **Y** forms very hard crystals which melt at 700 °C. Neither the solid nor the liquid conduct electricity. *(2 marks)*

d **Z** is a fairly dense, hard solid which can be flattened out if hammered. It melts at about 1100 °C and conducts electricity both as a solid and a liquid. *(2 marks)*

8 a Suggest a reason why 'nano' catalysts might work better than ordinary catalysts. *(1 mark)*

b How could 'nano' catalysts help to clean up the environment? *(1 mark)*

c What are nanobots, and how could they be usefully deployed? *(2 marks)*

d Why are some people worried about the development of nanobots? *(1 mark)*

e Read this newspaper article about nanotechnology and the future, then answer the questions below.

Nanotechnology is the new technology of the very small – down to the size of atoms and molecules. Scientists are getting very excited about it because it opens up a whole range of new possibilities, such as improved medicines or solar cells. But it has also scared a lot of people who worry about what could happen if these new technologies got out of control.

In 2005 a 'Citizen's jury' was set up so that 'ordinary people' could discuss the benefits and dangers of nanotechnology. Their report, issued in September 2005, was quite upbeat. They recommended that the government should support research into health, environmental and energy production uses of nanotechnology. They thought that these new technologies could give direct benefit and would also help to create jobs and keep the economy in good shape.

But they also recognised that careful testing would be necessary before the technology is approved. The jury recommended that the government should be more open about its plans for the future, and should keep people informed.

  i Why was the 'Citizen's jury' set up? *(1 mark)*

  ii What are the **three** areas where the jury thought that nanotechnology could be most helpful? *(3 marks)*

  iii The newspapers had been full of scare stories about nanotechnology, yet the jury's positive report was given very little space. Suggest a reason for this. *(1 mark)*

9 a What are the **two** particles found in the nucleus of an atom? *(2 marks)*

b What is the mass of each particle (compared to a hydrogen atom)? *(2 marks)*

c How many of each of these particles are there in the nucleus of

  i $^{23}_{11}Na$? **ii** $^{108}_{47}Ag$? **iii** $^{238}_{92}U$? *(6 marks)*

d $^{20}_{10}Ne$ and $^{21}_{10}Ne$ are both *isotopes* of neon. What does *isotopes* mean? *(2 marks)*

e Copy and complete the table below.

	Atomic number	Mass number	Number of protons	Number of neutrons
carbon $^{12}_{6}C$	6	12	6	
nitrogen $^{14}_{7}N$	7		7	7

(2 marks)

f What is the atomic number of an isotope $^{14}_{6}X$? *(1 mark)*

g Is X an isotope of carbon or nitrogen? Explain your answer. *(2 marks)*

h The relative atomic mass for hydrogen is 1.008 not 1. Why is this? (What is it compared with?) *(1 mark)*

10 Copper has two oxides, CuO and Cu_2O (Cu = 64, O = 16).

a What would the simple Cu:O mass ratio be in
  i CuO?   **ii** Cu_2O?   *(2 marks)*

b What would the percentage of copper by mass in each oxide be?   *(1 mark)*

c A 2 g sample of copper was heated in air until it had completely reacted with oxygen. This gave 2.5 g of copper oxide.
  i How much oxygen reacted with the copper?   *(1 mark)*
  ii What is the mass ratio of copper to oxygen in this oxide?   *(1 mark)*
  iii Is this oxide CuO or Cu_2O?   *(1 mark)*

11 Here are some unbalanced chemical equations. For each one
  i write it out as a word equation
  ii copy and balance the symbol equation.

a _____ $H_2 + O_2 \rightarrow 2H_2O$ *(2 marks)*

b Zn + _____ $HCl \rightarrow ZnCl_2 + H_2$ *(2 marks)*

c _____ $LiOH + H_2SO_4 \rightarrow Li_2SO_4 + 2H_2O$ *(2 marks)*

d $CaCO_3$ + _____ $HCl \rightarrow CaCl_2 + H_2O + CO_2$ *(2 marks)*

e _____ K + _____ $H_2O \rightarrow 2KOH + H_2$ *(3 marks)*

f CH_4 + _____ $O_2 \rightarrow CO_2$ + _____ H_2O *(3 marks)*

12 What is the percentage of metal in each of these ores?

a galena, PbS_2 (Pb = 207, S = 32)   *(2 marks)*

b malachite, $CuCO_3$ (Cu = 64, C = 12, O = 16)   *(2 marks)*

c rutile, TiO_2 (Ti = 48)   *(2 marks)*

d cinnabar, HgS (Hg = 201)   *(2 marks)*

e barites, $BaSO_4$ (Ba = 137)   *(2 marks)*

13 a How many particles would you have in each of the following cases?
  i carbon dioxide molecules in 44 g of carbon dioxide   *(2 marks)*
  ii copper atoms in 32 g of copper metal   *(2 marks)*
  iii water molecules in 180 g of water   *(2 marks)*

b $CaCO_3 \rightarrow CaO + CO_2$
  i What is the formula mass of calcium carbonate? (Ca = 40)   *(2 marks)*
  ii How many moles of calcium carbonate would you have if you had 150 g of it?   *(1 mark)*
  iii How many moles of carbon dioxide would this produce if you heated it until it had reacted completely?   *(1 mark)*
  iv What would the volume of that gas be at atmospheric pressure and room temperature?   *(1 mark)*

14 Petrol contains some heptane, C_7H_{16} (as well as octane). The following reaction occurs when heptane burns:

$C_7H_{16} + 11O_2 \rightarrow 7CO_2 + 8H_2O$

a How many moles of oxygen are needed for every mole of heptane?   *(1 mark)*

b What is the formula mass of heptane?   *(2 marks)*

c One litre of heptane has a mass of about 700 g. How many moles of heptane do you get per litre?   *(1 mark)*

d How many moles of oxygen would you need to burn 1 litre of heptane?   *(2 marks)*

e What volume would that have (at atmospheric pressure and room temperature)?   *(2 marks)*

f Air is one-fifth oxygen. What volume of air (at atmospheric pressure and room temperature) would you need to burn 1 litre of heptane?   *(2 marks)*

g How many moles of carbon dioxide do you get when you burn 1 mole of heptane?   *(2 marks)*

h How many moles of carbon dioxide do you get when you burn 1 litre of heptane?   *(2 marks)*

i What volume would that have (at atmospheric pressure and room temperature)?   *(2 marks)*

15 Aluminium is made by the electrolysis of bauxite, Al_2O_3. The part equation is:

$Al_2O_3 \rightarrow 2Al$ (+ oxygen)

a What is the formula mass of bauxite? (Al = 27, O = 16)   *(2 marks)*

b What mass of aluminium would you expect to get from 1 mole of bauxite? (the theoretical yield) *(2 marks)*

c In one real electrolytic cell, only 49 g of aluminium was actually produced per mole of bauxite (the actual yield). What was the percentage yield for this reaction? *(2 marks)*

d Suggest possible reasons why the percentage yield was not 100%. *(2 marks)*

16 Bromine water is made by adding bromine to water. Coloured bromine has a reversible reaction in water, which produces two colourless acids.

$$Br_2(aq) + H_2O \rightleftharpoons HBr + HOBr$$
   brown   colourless

This reaction reaches an equilibrium, so bromine water has a mixture of all four substances and so is yellow-brown in colour.

a At equilibrium, has the reaction stopped? Explain your answer. *(2 marks)*

b This is a 'closed' system. If you add alkali, the acid particles are removed.
  i What has it now become? *(1 mark)*
  ii Which reaction, forward or back, would now go to completion? *(1 mark)*
  iii What would you *see* happen if you did this? *(1 mark)*

c Olive oil decolourises bromine water. Which substance is being removed from the system? *(1 mark)*

17 The diagram shows the main processes involved in the manufacture of ammonia.

nitrogen + hydrogen ⇌ ammonia

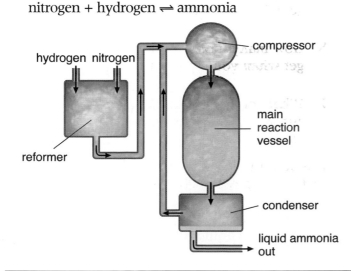

a Why doesn't all the nitrogen and hydrogen turn to ammonia in the main reaction vessel? *(1 mark)*

b How is the ammonia removed from the system so that the unused nitrogen and hydrogen may be recycled? *(1 mark)*

c What is the source of the nitrogen for this reaction? *(1 mark)*

d Copy and balance the equation for this reaction.
$$N_2 + \underline{\quad} H_2 \rightleftharpoons \underline{\quad} NH_3 \qquad \textit{(2 marks)}$$

e Calculate the percentage by mass of nitrogen in ammonia (NH_3). ($N = 14$, $H = 1$) *(2 marks)*

f What is the atom economy of this reaction? *(2 marks)*

g From the information so far, how do you view the Haber process in terms of sustainable development? *(2 marks)*

h The hydrogen is currently made from natural gas. Does that alter your view? Explain your answer. *(2 marks)*

18 Alan put a beaker containing 200 cm³ of 1 M hydrochloric acid and a watch glass with 1 g of marble chips onto an electric balance. He then zeroed the display on the balance. After this he tipped the limestone into the acid and put the empty watch glass back on the balance next to the beaker. He recorded the readings from the scale every 20 seconds.

Time (s)	20	40	60	80	100	120	140
Balance reading (g)	−0.11	−0.23	−0.33	−0.40	−0.43	−0.44	−0.44

a Alan held a drop of limewater on the end of a glass rod over the beaker and it turned milky. What gas was given off? *(1 mark)*

b Complete the equation.
$$CaCO_3 + 2HCl \rightarrow CaCl_2 + H_2O + \underline{\quad}$$
(1 mark)

c Explain why the reading on the balance scale dropped below zero. *(1 mark)*

d Plot a graph of Alan's results. Put time along the *x*-axis and mass loss in grams up the *y*-axis. Draw a 'best fit' curve. *(3 marks)*

e What would Alan have seen happening over the first minute or so? *(1 mark)*

f Suggest **two** ways in which Alan would have

known the reaction had finished after two minutes. *(2 marks)*

19 Petonelle used the same reaction and apparatus as Alan (question **18**) to investigate how the reaction rate changed with temperature. In her experiment she timed how long it took for the mass loss to reach –0.20 g. She repeated the experiment at different temperatures, to see how this affected the reaction.

Temperature (°C)	0	12	25	37	50
Time to reach –0.20 g (s)	150	72	37	19	10

a Suggest some safety precautions that would need to be taken for this experiment. How could the acid be heated safely, how could it be cooled? *(3 marks)*

b Plot a graph of Petonelle's results, with temperature along the horizontal. Draw a 'best fit' curve. *(3 marks)*

c Describe the pattern you see in words. How is 'the time it takes' changing as the temperature rises? *(1 mark)*

d What does that tell you about the way the speed of the reaction is changing as the temperature rises? *(1 mark)*

e Explain this effect in terms of what is happening at the particle level. *(2 marks)*

f Add a third row to the table and calculate the rate of the reaction at each temperature. (rate = 0.20 g /time taken). *(2 marks)*

g Plot a new graph of rate against temperature and draw the 'best fit' line. *(3 marks)*

h From your 'best fit' line, how much do you need to raise the temperature by to double the rate of this reaction? *(2 marks)*

i Petonelle actually repeated each experiment three times at each temperature. The figures shown are the average of the three. Why did she do this? *(2 marks)*

20 a Sparklers are made from tiny iron filings glued onto an iron wire. Explain why the iron filings burn but the iron wire does not. *(2 marks)*

b Rima wanted to find out which was more reactive, iron or nickel, so she put samples of each into tubes of 1 M hydrochloric acid, to see which bubbled the most. She had iron filings and nickel granules. Suggest why her results may not be reliable. *(2 marks)*

c Iron filings usually fizz faster in acid than iron nails. You might expect very fine iron dust to fizz even faster – but if it has settled out in a layer at the bottom that may not be the case.
  i Suggest a reason why. *(1 mark)*
  ii How could you easily get the iron dust to react faster again? Explain your answer. *(2 marks)*

d Iron acts as a catalyst in the Haber reaction.
  i Why does the iron have to be in very tiny pieces? *(1 mark)*
  ii Why does this iron have to be supported on a grid or mesh within the reacting gases? *(1 mark)*

21 In the old days before digital photography, people sometimes developed their own films at home using a developer chemical to react with the film and bring out the picture. This was bought in a concentrated form and diluted down with water before use. The time taken to develop the picture depends on both concentration and temperature, and may be worked out using this graph.

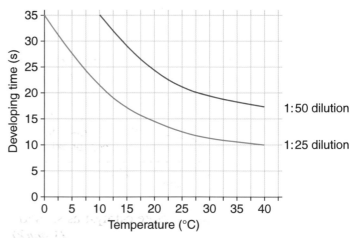

a What effect does temperature have on the time taken to develop the film? *(1 mark)*

b What effect does dilution have on the time taken to develop the film? Explain your answer in terms of the particles involved. *(2 marks)*

c How long should it take to develop a photograph at 27 °C if the developer has been diluted 1 : 25 with water? *(1 mark)*

d How long should it take to develop a

photograph at 27 °C if the developer has been diluted 1 : 50 with water? *(1 mark)*

22 Des was investigating the way hydrogen peroxide broke down when metal oxide powders were added. She had a flask containing hydrogen peroxide conncctcd to a 100 cm³ gas syringe. She added a spoonful of oxide, quickly replaced the bung and took a reading every 20 seconds.

a What factors, apart from which metal oxide she chooses, must Des keep constant if this is to be a fair test? *(2 marks)*

b Copy and complete the equation for this reaction.

$$2H_2O_2 \rightarrow 2H_2O + \underline{\hspace{2cm}}$$ *(1 mark)*

c The metal oxide acts as a catalyst for this reaction. Explain the term catalyst. *(2 marks)*

d Here is Des's graph for manganese dioxide, nickel oxide and copper oxide.

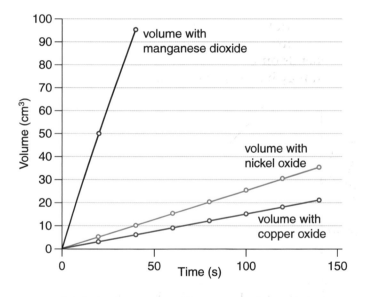

Which is the most effective catalyst as shown by these results? *(1 mark)*

e Calculate the reaction rate for each of the catalysts. *(3 marks)*

f Des thought she'd repeat the experiment with chopped liver as she had heard that it contained an enzyme that worked for this reaction. It reacted so fast that the bung popped out of the flask. She then repeated the experiment with cooked liver and the reaction was too slow to measure. Explain this result. *(2 marks)*

23 For each of the following reactions, state whether it is exothermic or endothermic. If it is exothermic, state whether it needs a 'kick-start' of energy or not.

a candle wax burning *(2 marks)*

b photosynthesis *(1 mark)*

c electrolysis of sodium chloride *(1 mark)*

d iron displacing copper from copper sulphate solution *(2 marks)*

e aluminium displacing iron from iron oxide powder *(2 marks)*

f baking soda dissolving in lemon juice *(1 mark)*

g caustic soda oven cleaner dissolving grease *(2 marks)*

h making sodium chloride from sodium hydroxide and hydrochloric acid *(2 marks)*

i gunpowder exploding *(2 marks)*

j paraffin burning in a jet enginc *(2 marks)*

k making copper oxide by heating copper carbonate *(1 mark)*

24 The following graphs were produced by a data-logging set-up using a temperature probe.

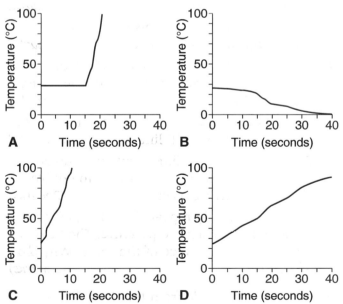

Which reaction below could have produced which graph shown above?

a solid sodium hydroxide dropped into hydrochloric acid *(1 mark)*

b a spoonful of baking soda dropped into a solution of citric acid *(1 mark)*

c iron filings dropped into copper sulphate solution *(1 mark)*

d a match head lit after 15 seconds *(1 mark)*

25 Bloggs and Co. are setting up a business to make chemical **Z**. When the solid chemical **X** is heated it decomposes to give two chemicals, **Y** and **Z**. This forward reaction is endothermic. Gas **Y** dissolves in water, gas **Z** does not. The reaction is shown by the equation: $X \rightleftharpoons Y + Z$

a What does the symbol $\rightleftharpoons$ tell you about the reaction? *(1 mark)*

b The chief chemist of the company has said that this reaction is likely to reach equilibrium with a 50% yield of **Y** and **Z** at 300 °C. Explain the terms equilibrium and yield. *(2 marks)*

c Mr Bloggs wants a better yield and thinks that changing the temperature might help. Should he raise or lower the temperature? Explain your answer. *(2 marks)*

d Suggest a way that the unwanted gas **Y** could be removed from the product mixture to give pure **Z**. *(2 marks)*

26 Lead iodide has the formula PbI_2. Molten lead iodide can be electrolysed using carbon electrodes as shown below.

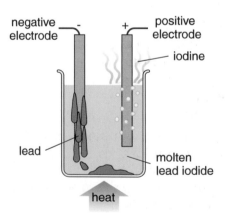

a Copy and complete these 'half equations' for the reactions at the electrodes:

at the negative electrode:
$Pb^{2+} + \_\_\_\_\_ e^- \rightarrow Pb$ *(1 mark)*

at the positive electrode:
$\_\_\_\_\_ I^- \rightarrow 2e^- + I_2$ *(1 mark)*

b What is the formula mass of lead iodide? (Pb = 207, I = 127)

c How many moles of I_2 molecules will you get from 1 mole of lead iodide? *(2 marks)*

d If the experiment produced 2.07 g of lead, how many grams of iodine would you get? *(2 marks)*

e Molten lead chloride ($PbCl_2$) can also be electrolysed in this way. Write half equations for the reactions at each of the electrodes. *(2 marks)*

f What is the formula mass for lead chloride? (Cl = 35.5) *(2 marks)*

g How many moles of Cl_2 molecules will you get from 1 mole of lead chloride? *(1 mark)*

h What volume of chlorine gas would you get from 2.78 g of lead chloride? (At room temperature and atmospheric pressure.) *(2 marks)*

27 Look at this diagram, showing the products that can be made from sodium chloride, common salt.

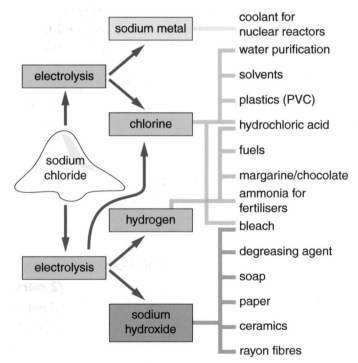

a Household bleach is made by bubbling chlorine through sodium hydroxide solution. Draw a flow chart for the production of bleach from salt. *(2 marks)*

b Explain why metallic sodium cannot be

made by the electrolysis of sodium chloride solution. *(2 marks)*

c At which electrode does hydrogen form during the electrolysis of sodium chloride solution? *(1 mark)*

d Copy and complete the half equation for this reaction:

_____ $H^+ + 2e^- \rightarrow H_2$ *(1 mark)*

e Is the reaction in question **d** oxidation or reduction? *(1 mark)*

f Why is the electrolysis of sodium chloride important for human health? *(1 mark)*

g What is the atom economy for the electrolysis of sodium chloride? (What proportion of the products are used?) *(1 mark)*

h What is the problem with this electrolysis reaction in terms of sustainable development? *(1 mark)*

i Suggest how this problem might be overcome. *(1 mark)*

28 Silver nitrate ($AgNO_3$) dissolves in water, releasing its Ag^+ silver ions. Silver nitrate solution can be used in an electrolytic cell to silver plate metal objects such as nickel spoons.

a To which electrode would you have to connect a nickel spoon to silver plate it? *(1 mark)*

b Write the half equation for the formation of silver from the silver ion at this electrode. *(1 mark)*

c Is **b** oxidation or reduction? *(1 mark)*

d If you simply used silver nitrate solution, what would happen to its concentration as the electrolysis went on? *(1 mark)*

29 Add state symbols to each of these equations.

a $2H_2 + O_2 \rightarrow 2H_2O$ *(2 marks)*
b $Zn + 2HCl \rightarrow ZnCl_2 + H_2$ *(2 marks)*
c $2LiOH + H_2SO_4 \rightarrow Li_2SO_4 + 2H_2O$ *(2 marks)*
d $CaCO_3 + 2HCl \rightarrow CaCl_2 + H_2O + CO_2$ *(2 marks)*
e $2K + 2H_2O \rightarrow 2KOH + H_2$ *(2 marks)*
f $CH_4 + 2O_2 \rightarrow CO_2 + 2H_2O$ *(2 marks)*

30 Sodium chloride and sodium carbonate are soluble in water. Calcium chloride is soluble in water, but calcium carbonate is not.

a What ions are there in a solution of sodium carbonate? *(2 marks)*

b What ions are there in a solution of calcium chloride? *(2 marks)*

c Explain what will happen if the two solutions in questions **a** and **b** are mixed. *(2 marks)*

d Write a word equation for this. *(1 mark)*

e Many areas of the country have 'hard water' – water with calcium salts dissolved in it. Soap contains a soluble chemical called sodium stearate. Calcium stearate is insoluble. Explain what will happen if you use soap in a hard water area. *(2 marks)*

f Write a word equation for question **e**. *(2 marks)*

The calcium stearate that forms is called scum. Scum is greasy and unpleasant to have in your bath, so many people add 'bath salts' to the water to 'soften' the water and stop scum forming. Bath salts are simply perfumed and coloured sodium carbonate crystals.

g Explain what happens to the calcium ions when bath salts are added. *(2 marks)*

h How does this stop the scum from forming? *(2 marks)*

31 Acid spills can be very dangerous. Acid spills in the laboratory are cleaned up using sodium hydrogencarbonate ($NaHCO_3$).

a What is the reaction that 'cancels out an acid' called? *(1 mark)*

b If 500 ml of 1 M hydrochloric acid was spilt, it could be neutralised by exactly the right amount of caustic soda (sodium hydroxide). Suggest two reasons why this option is not used. (Hint: is this reaction endothermic or exothermic?) *(2 marks)*

c In practice, sodium hydrogencarbonate is liberally sprinkled over acid spills.
   i What would you see happening? *(1 mark)*
   ii How would you know when all the acid was neutralised? *(1 mark)*

d Copy and complete the equation for the reaction described in question **d**.

$HCl + NaHCO_3 \rightarrow$ _____ $+ H_2O + CO_2$ *(1 mark)*

e Why is it not a problem if you use too much sodium hydrogencarbonate? *(1 mark)*

The Periodic Table

Now chemists all over the world use the same symbols for the elements, work out formulae the same way and know what each other means by relative atomic mass. You see the same **Periodic Table** in laboratories worldwide.

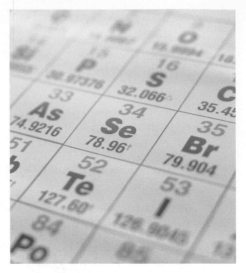

Chemical chaos

It was not always so. Before 1860 chemists knew of 59 elements and had agreed names and symbols for most of them. But they could not agree how to put symbols together to make formulae. There were 15 different formulae for ethanoic acid, the acid in vinegar, and none of them were wrong!

Chemists had not agreed how to measure the mass of atoms. They did not even agree on what was an atom and what was a molecule! There were two different scales, one with oxygen at 16 and one with oxygen at 32. Some chemists worked with one scale, some with the other, without making it clear which scale they were using. It was like measuring with metre sticks of different lengths.

On 3 September 1860 the International Congress of Chemists began. At the first ever international meeting of scientists, held in Karlsruhe, Germany, 140 scientists from 12 countries tried to bring some order to the chaos. They debated for three days.

A passionate debate

There are no photographs of the meeting: photography was a very new invention. We have some photos of those who attended, taken years later. They are in shades of grey, showing middle-aged and elderly men in formal, stiff poses. Were the proceedings as stuffy as the participants appear?

Absolutely not. Six of the scientists kept notes. That's how we know that the meeting was lively and that the scientists were passionate in their arguments. Those who spoke were determined to make their points and convince their audience, none more than Stanislao Cannizzaro, a 34-year-old professor of chemistry from Italy.

A man with a mission

Stanislao Cannizzaro never made a great scientific discovery, yet he is considered to be one of the most important chemists of all time. He realised how important it was to settle on one system for measuring the relative mass of an atom, called the **atomic weight**. Sort this out and chemical formulae would follow.

▲ Cannizzaro – the great communicator.

Cannizzaro was a great communicator. He dominated the final day of the meeting, speaking for longer than anyone else. During his clear and consistent argument the now familiar formulae for substances emerged: H_2O for water, CH_4 for methane, O_2 for oxygen gas, CH_3CO_2H for ethanoic acid. He even had a handout for each chemist to take home!

Officially, the meeting did not reach agreement: they decided to think more on the matter. In practice, the argument was won: oxygen was given atomic weight 16 on the scale, which gave hydrogen an atomic weight of 1. These chemists, and all those who came after, adopted this scale and this system for expressing formulae. Now they were all using a metre stick of the same length, developing and discussing ideas became much easier.

Think about what you will find out in this section

Where did the Periodic Table come from?	How does the modern Periodic Table relate to atoms?
Why are some elements so lively when their compounds are well behaved?	Why are all transition elements so alike?

Mendeleev

By 1860, 59 elements had been discovered and chemists struggled to make sense of them. Many ideas were suggested, but the first to include all the known elements was proposed by Dmitri Mendeleev in 1869.

Mendeleev had attended the International Congress in 1860. Cannizzaro had convinced him that atomic weight was crucial. Mendeleev put the elements in order of their atomic weights and then tried to line them up so that similar elements were in horizontal rows. He found that elements showed a repeating pattern or **periodicity**. His first Periodic Table, published in 1869, is shown here. It shows the elements' symbols and atomic weights. The colours have been added to show the modern groups, so that you can compare Mendeleev's version to the Periodic Table we use today (see section 1.2).

Filling the gaps

When you consider the inaccuracy of some of Mendeleev's information, his achievement is even more impressive. Didymium, Di, was not an element at all, and all the elements in blue had wildly inaccurate atomic weights.

Mendeleev left gaps in his table to maintain the periodicity (the repeating pattern). He predicted that elements would be discovered that fitted into these gaps. He even predicted their properties. For example, he predicted the discovery of an element between silicon, Si, and tin, Sn. The element germanium, Ge, was discovered 17 years later and had all the properties that Mendeleev had predicted.

			Ti = 50	Zr = 90	GAP = 180
			V = 51	Nb = 94	Ta = 182
			Cr = 52	Mo = 96	W = 186
			Mn = 55	Rh = 104.4	Pt = 197.4
			Fe = 56	Ru = 104.4	Ir = 198
			Ni = Co = 59	Pd = 106.6	Os = 199
H = 1			Cu = 63.4	Ag = 108	Hg = 200
	Be = 9.4	Mg = 24	Zn = 65.2	Cd = 112	
	B = 11	Al = 27.4	GAP = 68	U = 116	Au = 197?
	C = 12	Si = 28	GAP = 70	Sn = 118	
	N = 14	P = 31	As = 75	Sb = 122	Bi = 210?
	O = 16	S = 32	Se = 79.4	Te = 128?	
	F = 19	Cl = 35.5	Br = 80	I = 127	
Li = 7	Na = 23	K = 39	Rb = 85.4	Cs = 133	Tl = 204
		Ca = 40	Sr = 87.6	Ba = 137	Pb = 207
		GAP = 45	Ce = 92		
		?Er = 56	La = 94		
		?Y = 60	Di = 95		
		?In = 75.6	Th = 118?		

Groups in the modern Periodic Table

Group 1	Group 2	Group 3	Group 4
Group 5	Group 6	Group 7	Transition elements

▲ Mendeleev's first Periodic Table, colour-coded to show the modern groups.

Questions

a Make two lists, one of the similarities of Mendeleev's table to the modern Periodic Table and one of the differences.

b The correct relative atomic masses for the elements in blue are as follows: In = 114.8, U = 238.0, Er = 167.3, Y = 88.9, Ce = 140.1, La = 138.9, Th = 232.0. Use this, and the fact that Di is not an element, to produce a 'next version' of Mendeleev's table.

Was Mendeleev first?

Other scientists had worked on the same problem. Mendeleev himself, writing in 1889, recognised the contribution of a number of scientists, including Alexandre de Chancourtois (a French geologist), John Newlands (a British chemist) and Julius Meyer (a German chemist). These three scientists also ordered elements according to their atomic weight, and then looked for repeating (periodic) patterns.

De Chancourtois plotted the elements on the surface of a cylinder, so that similar elements were above each other in a helix pattern. Newlands proposed a 'Law of Octaves', shown below, based on the same repeating pattern in properties that Mendeleev reported. Both these scientists reported their findings in 1863, six years before Mendeleev, but neither of them convinced their fellow scientists. De Chancourtois used geological rather than chemical language in his report and did not include a diagram. This made his ideas difficult to understand. Newlands made an analogy between his 'Law of Octaves' and a musical octave. This analogy was unpopular with his fellow scientists, who then failed to see the good science behind his ideas.

▲ Dmitri Mendeleev.

H	F	Cl	Co & Ni	Br	Pd	I		Pt & Ir
Li	Na	K	Cu	Rb	Ag	Cs		Os
Be	Mg	Ca	Zn	Sr	Cd	Ba	& V	Hg
B	Al	Cr	Y	Ce & La	U	Ta		Tl
C	Si	Ti	In	Zr	Sn	W		Pb
N	P	Mn	As	Di	& Mo	Sb	Nb	Bi
O	S	Fe	Se	Ro & Ru	Te	Au		Th

Groups in the modern Periodic Table

- Group 1
- Group 2
- Group 3
- Group 4
- Group 5
- Group 6
- Group 7
- Transition elements

▲ Newlands' Law of Octaves, colour-coded to show the modern groups.

Question

c Newlands had no gaps in his 'table'. Why was this a problem?

Julius Meyer was working on a Periodic Table at the same time as Mendeleev. He had put the elements into groups and left gaps for undiscovered elements. Unfortunately for Meyer, he published his ideas in 1870, while Mendeleev published his in 1869. In most scientists' minds, Mendeleev had got there first and was given credit for the idea.

Question

d In your opinion, is publication date the best way to decide 'ownership' of an idea? What other evidence could be used?

Key points

- Dmitri Mendeleev arranged the known elements in order of atomic weight. Elements with similar chemical properties appeared at regular intervals in his first Periodic Table published in 1869.
- Mendeleev realised that some elements existed but had not been discovered. He left spaces for them. Later, elements were discovered that fitted in the gaps.
- Mendeleev's Periodic Table was not the first, or the only, proposal. His was the first convincing version, so he was given the credit.

Same but different

Our Periodic Table is different from Mendeleev's in a number of ways, reflecting the discoveries that have been made during the last 140 years.

Group

Period	1	2		3	4	5	6	7	8	9	10	11	12	3	4	5	6	7	0
1						1 H													2 He
2	3 Li	4 Be												5 B	6 C	7 N	8 O	9 F	10 Ne
3	11 Na	12 Mg												13 Al	14 Si	15 P	16 S	17 Cl	18 Ar
4	19 K	20 Ca		21 Sc	22 Ti	23 V	24 Cr	25 Mn	26 Fe	27 Co	28 Ni	29 Cu	30 Zn	31 Ga	32 Ge	33 As	34 Se	35 Br	36 Kr
5	37 Rb	38 Sr		39 Y	40 Zr	41 Nb	42 Mo	43 Tc	44 Ru	45 Rh	46 Pd	47 Ag	48 Cd	49 In	50 Sn	51 Sb	52 Te	53 I	54 Xe
6	55 Cs	56 Ba	*	71 Lu	72 Hf	73 Ta	74 W	75 Re	76 Os	77 Ir	78 Pt	79 Au	80 Hg	81 Tl	82 Pb	83 Bi	84 Po	85 At	86 Rn
7	87 Fr	88 Ra	**	103 Lr	104 Rf	105 Db	106 Sg	107 Bh	108 Hs	109 Mt	110 Ds	111 Rg	112 Uub	113 Uut	114 Uuq	115 Uup	116 Uuh		

*Lanthanoids	*	57 La	58 Ce	59 Pr	60 Nd	61 Pm	62 Sm	63 Eu	64 Gd	65 Tb	66 Dy	67 Ho	68 Er	69 Tm	70 Yb
**Actinoids	**	89 Ac	90 Th	91 Pa	92 U	93 Np	94 Pu	95 Am	96 Cm	97 Bk	98 Cf	99 Es	100 Fm	101 Md	102 No

▲ The modern Periodic Table.

New elements

Many new elements have been discovered, as well as the four elements Mendeleev predicted, bringing the current total to 116. The most important discovery was of a whole new group, the noble gases or Group 0, at the end of the nineteenth century.

> **Question**
>
> **a** Scientists thought they had discovered element 118 but no other laboratories could reproduce their results. Why is confirmation by another group of scientists important?

A new order

The discovery of the electron, the proton and the neutron in the early twentieth century led to scientists working out the structure of atoms. In 1913 the physicist Henry Moseley arranged the elements according to how many protons they had, introducing the idea of **atomic number**.

Arranging the elements according to atomic weight had caused problems. For example, argon, Ar, has an atomic weight of 39.9 and potassium, K, has an atomic weight of 39.1. Using atomic weights, potassium should come before argon. This would put potassium in Group 0 and argon in Group 1. Once you start working with atomic number rather than atomic weight, this problem vanishes. Argon is atomic number 18 and potassium is atomic number 19, placing them in the correct groups.

Question

b Argon is a colourless, inert gas and potassium is a soft, reactive metal. Explain why argon does not belong in Group 1 and potassium does not belong in Group 0.

Electrons rule

In fact, as you know, the modern Periodic Table not only reflects patterns in the properties of elements, it is also a summary of the number of electrons in an atom and the way they are arranged. Period 1 has only two elements because there are places for two electrons in the lowest energy level. Period 2 has eight elements, because there are places for eight electrons in the second energy level. Elements in a group have the same number of electrons in their outermost energy level. This is why they react in similar ways: it is the outer electrons that do the reacting!

Question

c Group 0 was originally called Group 8. Suggest why the name was changed.

Look at Period 4. Potassium, K, has one electron in the fourth energy level, calcium, Ca, has two, gallium, Ga, has three and so on until you reach Krypton, Kr, that has eight electrons in the fourth energy level. So what is the story for the **transition elements** (shown in grey) that come between calcium and gallium?

The answer is that the third energy level has another 10 places to fill. These places are filled after the first two places in the fourth energy level. The transition elements in Period 4 each have one more electron in places 9–18 of energy level 3. Likewise, the transition elements in Period 5 each have one more electron in places 9–18 of energy level 4. This order of 'electron place filling' is summed up in the diagram. Follow the arrows to see the order in which the places in each energy level are filled. The colours correspond to the groups in the Periodic Table.

Question

d Calcium has electronic structure 2.8.8.2. Give the electronic structures of:
(i) scandium, Sc (atomic number 21)
(ii) zinc, Zn (atomic number 30).

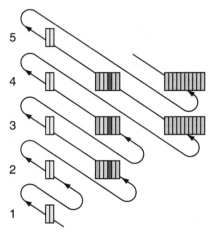
▲ The order in which the electrons fill the energy levels.

Key points

- The modern Periodic Table is based on atomic number rather than atomic weight.
- Each Group (column) represents a particular arrangement of the outer electrons of the atom. That is why elements with similar properties appear there.
- As you move downwards through the Periods (rows) additional, higher energy levels are filled with electrons. The pattern of elements reflects the energy level being filled.

Group 1: Lively metals

Uncommon metals

The elements of Group 1, the **alkali metals**, have all the familiar metallic properties: they are shiny, they conduct electricity and thermal energy, they can be beaten into sheets and pulled into wire. Even so, you would not want to use these metals as jewellery, electrical wiring or for wrapping your food!

Soft and floaty

Group 1 metals are soft. You can cut lithium with a knife and the elements get even softer as you go down the group. Group 1 metals are much less dense than most metals. They also have low melting and boiling points: caesium would melt on a very hot day. Look at the table. It shows the densities, melting and boiling points of the Group 1 elements (in bold) and some other metals.

Element	Density (g/cm³)	Melting point (°C)	Boiling point (°C)
Lithium	0.53	181	1342
Sodium	0.97	98	883
Potassium	0.86	63	760
Rubidium	1.53	39	686
Caesium	1.88	29	669
Aluminium	2.70	660	2467
Copper	8.92	1083	2567
Iron	7.86	1535	2759
Gold	18.88	1064	3080

Question

a Which of the Group 1 metals shown in the table
(i) is the densest? (ii) is the softest? (iii) needs the most thermal energy to melt? (iv) would float on water?

Too reactive

Look at the photos of lithium, Li, sodium, Na, potassium, K, rubidium, Rb, and caesium, Cs. The lumps of lithium, sodium and potassium are all tarnished on the surface. This is because they react quickly with the oxygen in the air. Even a newly cut surface is only briefly shiny, tarnishing quickly. You have to store these metals under oil.

You can see that the rubidium and the caesium are in glass tubes. These glass tubes contain an inert gas instead of air. This is because the metals would react so quickly with the oxygen in the air that there would be a risk of fire.

Question

b Suggest an inert gas that could be used in the glass tubes.

2.1 Li

2.8.1 Na

2.8.8.1 K

2.8.18.8.1 Rb

2.8.18.18.8.1 Cs

▲ Group 1 elements.

When you put lithium in water you see bubbles. The lithium reacts with the water, making hydrogen, and the temperature of the water increases. If you react potassium with water, much more thermal energy is released more quickly: the hydrogen bursts into flames. The flame is lilac in colour because of the potassium. When caesium is put in water it explodes!

▲ Lithium bubbles in water.

Question

c *Predict what will happen when you put rubidium in water.*

Losing electrons

So why do the Group 1 elements become more reactive as you move down the group? You already know that the alkali metals react with non-metal elements to make ionic compounds. For example, lithium reacts with oxygen to make lithium oxide. Lithium oxide contains Li$^+$ ions. Each lithium atom has lost its single, outer electron. You may recognise this as an **oxidation** reaction.

▲ Potassium bursts into flame in water.

The same happens when a sodium atom reacts to make a sodium ion, but the single electron comes from the third energy level (third electron shell) rather than the second. It is easier to remove an electron from the third energy level than the second, so sodium is more reactive than lithium. It is oxidised more easily.

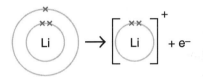
▲ Formation of a lithium ion.

Question

d *Draw a diagram showing a sodium atom making a sodium ion.*

The trend continues down the group. By the time you reach caesium the outer electron is in the sixth energy level and is even more easily lost, so caesium is more reactive than any of the alkali metals above it in Group 1.

The table below shows the energy needed to remove the outer electron from the same number of Group 1 atoms of each type. These measurements are made with the element as a gas.

Element	Energy to remove the outer electron from one mole of atoms (kJ)
Lithium	520
Sodium	496
Potassium	419
Rubidium	403

Question

e *What is the relationship between the energy needed to remove the outer electron from the atom and the reactivity of the element?*

Key points

- Group 1 of the Periodic Table consists of soft metals so reactive that they have to be stored under oil or gas.
- Uncover a Group 1 metal and it reacts immediately with the oxygen in the air. For the more reactive Group 1 metals, a fire could result.
- Going down the group, it takes less energy to remove the outermost electron from the atom and the elements are progressively more reactive.

Group 1: Boring compounds

Unlike the element

Group 1 elements are metallic, fizzy and flashy. In contrast, Group 1 compounds are white solids that dissolve calmly in water to make colourless solutions.

Give and take

Group 1 metals react with non-metals to make ionic compounds. All members of the group react in a similar way, because they all have one electron in their outermost energy level. For example, lithium reacts with oxygen to make lithium oxide:

lithium + oxygen → lithium oxide
$$4Li + O_2 → 2Li_2O$$

The atoms of lithium lose their outer electron to form lithium ions, Li^+. The lithium atoms are oxidised. The atoms of oxygen gain two electrons each to form oxide ions, O^{2-}. This is summed up in the diagram.

▲ Sodium chloride, a typical Group 1 compound.

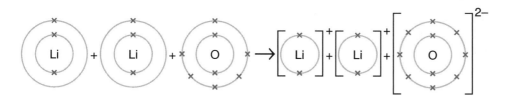

▲ Ionisation of lithium and oxygen atoms.

As you need two lithium atoms to provide enough electrons for one oxide ion, the ratio of reacting atoms is 2Li:1O, and the compound formed has the formula Li_2O.

Question

a Write balanced equations for sodium, potassium, rubidium and caesium reacting with oxygen.

A lot of energy is given out when lithium burns. You end up with ions bonded in ionic compounds. These compounds are a lot less reactive than the elements that made them. We say they are more **stable**.

Wanted or not

Alkali metals are so reactive that they even react with stable compounds, like water, producing an alkaline solution. This is why they are called alkali metals. You may remember studying **displacement** reactions. A more reactive metal 'pushes' a less reactive metal out of its compound.

Alkali metals displace hydrogen from water, in the same way. Look at the table. It shows a reactivity series that includes carbon and hydrogen as well as many metals. Alkali metals displace hydrogen in more

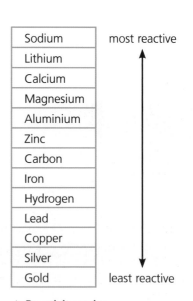

Sodium	most reactive
Lithium	
Calcium	
Magnesium	
Aluminium	
Zinc	
Carbon	
Iron	
Hydrogen	
Lead	
Copper	
Silver	
Gold	least reactive

▲ Reactivity series.

vigorous reactions. Look back at section 1.3 to see the reactions between potassium and water and between lithium and water.

Question

b Rewrite the reactivity series shown to include the other alkali metals.

The equation for the reaction between potassium and water is shown below:

potassium + water → potassium hydroxide + hydrogen

$$2K \quad + \quad 2H_2O \rightarrow \quad 2KOH \quad + \quad H_2$$

The diagram below shows the 'before' and 'after' of this reaction. The potassium atoms end up as more stable potassium ions. Some of the hydrogen atoms end up as less stable hydrogen gas, rather than the more stable compound, water.

It's all in a name

The reaction between potassium and water forms potassium hydroxide, an alkali, which dissolves in the water. If you add universal indicator to the water during this reaction, it turns from green (neutral) to dark purple (very alkaline). All the alkali metals react with water in the same way.

▲ Sodium reacts with water to make an alkaline solution of sodium hydroxide.

Question

c Write balanced symbol equations for the reactions of lithium and of rubidium with water.

Key points

- The Group 1 metals react readily even with stable compounds. They displace less reactive metals to form very stable ionic compounds, white solids that dissolve in water.
- When they react with water, they displace hydrogen in very vigorous reactions. The result is an alkaline solution.

The tiger of chemistry

Some materials are more difficult to work with than others. Group 7, the **halogens**, are some of the most dangerous. Fluorine was known as the tiger of chemistry. This pale yellow gas is so incredibly reactive that chemists had to fight to keep it contained – it reacted with almost everything it touched, including human flesh. Of the seven scientists who tried to work on it during the nineteenth century, three died, one almost blew himself up, and three were poisoned but recovered. Henri Moisson, one of those who recovered, was credited with its discovery in 1886. He used an apparatus made of platinum to isolate and study the gas.

> **Question**
>
> **a** Why was Henri Moisson's apparatus made of platinum?

Poison gas but safe water

Chlorine, the next member of Group 7, is less reactive than fluorine but still dangerous. This yellowish-green gas was one of the poison gases used by both sides during World War I. The drawing is a German soldier's impression of a poison gas attack. Chlorine destroyed the delicate tissues of the respiratory system: many soldiers died and others had permanent lung damage. After the war, there was an international agreement that poison gas would never be used in war again.

▲ Storm Troops Advance under Gas Attack drawn by Otto Dix, a German artist, in 1924.

> **Question**
>
> **b** Chlorine is used at low concentrations to kill bacteria in water, sterilising it. Explain why chlorine is used to do this rather than fluorine.

The also-rans

Bromine, the next member of Group 7, is a red-brown liquid at room temperature but evaporates quickly, making a red-brown gas. It is less reactive, and therefore less dangerous, than fluorine and chlorine, but still burns the skin on contact.

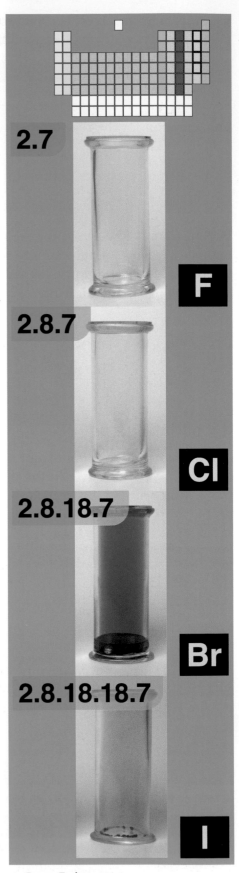

2.7   **F**

2.8.7   **Cl**

2.8.18.7   **Br**

2.8.18.18.7   **I**

▲ Group 7 elements.

Iodine is a grey solid at room temperature but it 'sublimes', giving off a violet vapour with an irritating smell. Although iodine can be poisonous at high concentrations, it is used at very low concentrations in antiseptic ointments.

Gas, liquid or solid?

The Group 7 elements are non-metals. Pairs of atoms are bonded as molecules. The diagram shows the electronic structure of a fluorine molecule, F_2.

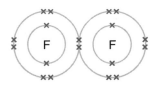
▲ A fluorine molecule.

Question

c Draw a diagram to show the electronic structure of a chlorine molecule.

Fluorine and chlorine are gases at room temperature, bromine is a liquid and iodine is a solid. This suggests that their boiling points increase down the group, and this is confirmed by the data in the table.

Element	Melting point (°C)	Boiling point (°C)
Fluorine	−220	−188
Chlorine	−101	−35
Bromine	−7	59
Iodine	114	184

Question

d (i) What is the relationship between the boiling points of the halogens and the size of their molecules?
(ii) Explain this relationship. (Hint: What weak forces exist between the molecules?)

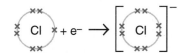
▲ Making a chloride ion.

Decreasing reactivity

When the Group 7 elements react they often form ionic compounds. For example, when chlorine reacts with sodium it makes sodium chloride, and when bromine reacts with iron it makes iron bromide. **Halide** ions (fluoride, chloride, bromide or iodide) are made during this reaction. The diagram shows the changes to the arrangement of outer electrons when a chloride ion is made from a chlorine atom. You may recognise this as a **reduction** reaction, as an electron has been gained.

In Group 7 the reactivity of the elements decreases as you go down the group. This is in contrast to Group 1, where the reactivity increases as you go down the group. The table shows the energy released when this change takes place. As you can see, more energy is released when you reduce a chlorine atom than when you reduce a bromine atom. This difference is also seen between bromine and iodine. For these three elements, the lower the energy level receiving the electron, the less energy released. This partly explains the decrease in reactivity from chlorine to iodine.

But it cannot be the whole story. Fluorine is much more reactive than chlorine, but less energy is released when you add an electron to a fluorine atom than when you add an electron to a chlorine atom. There must be other steps on the way to making fluoride compounds that release energy, making fluorine more reactive than chlorine.

Element	Energy released when one mole of electrons is added to one mole of halogen atoms, making one mole of halide ions (kJ)
Fluorine	−328
Chlorine	−349
Bromine	−325
Iodine	−295

Key points

- Halogens, in Group 7 of the Periodic Table, are very reactive non-metals that often form ionic compounds.
- The reactivity of the halogens decreases as you go down the group.
- Atoms bond in pairs, forming molecules in which one outer electron from each atom is shared.

Group 7: Compounds

Compounds galore

Fluorine is the most reactive of all the elements. It makes water burn and eats through glass. It even reacts with some noble gases. As you can imagine, this means that fluorine forms a huge variety of compounds. The less reactive halogens do not make as many, but the range is still impressive.

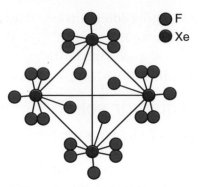

▲ Xenon hexafluoride, XeF_6.

> **Question**
>
> **a** The diagram shows the structure of xenon hexafluoride. Why was it a surprise when compounds between xenon and fluorine were first made in 1962?

Give and take

Given a chance, halogens will form halide ions. This needs electrons. Metals have electrons to spare, so the two elements react. Magnesium is a reactive metal found in Group 2 of the Periodic Table, so it has two outer electrons to give. Chlorine is in Group 7, so each chlorine atom has seven outer electrons and needs one more to make a full outer shell. Each magnesium atom can provide the electrons for two chlorine atoms. The magnesium is oxidised, making Mg^{2+} ions. The chlorine is reduced, making Cl^- ions. These ions bond together, making an ionic compound, a salt, called magnesium chloride, formula $MgCl_2$. The overall equation is as follows:

$$Mg + Cl_2 \rightarrow MgCl_2$$

> **Question**
>
> **b** Think about the reaction between aluminium and fluorine.
> (i) What ions will be made?
> (ii) What is the name and formula of the salt made?
> (iii) Write a balanced equation for the reaction.
> (iv) Which element was oxidised?
> (v) Which element was reduced?

Grab those electrons

You are familiar with displacement reactions. A more reactive metal will displace a less reactive metal from its compound. A similar reaction happens with the halogens. The more reactive halogen will displace the less reactive halogen from its compound. The photos show the reactants and the products when chlorine gas is bubbled through potassium bromide solution.

You will recall that the halogens become less reactive as you move down Group 7. In the battle to gain electrons and make halide ions, chlorine wins over bromine. The chlorine is reduced. The bromide ions are oxidised.

$$Cl_2 + 2Br^- \rightarrow 2Cl^- + Br_2$$

▲ Chlorine displaces bromine.

Question

c *Predict the products when bromine gas is bubbled into*
(i) a solution of potassium iodide, and
(ii) a solution of potassium chloride.
(iii) Write equations for any reactions.

Sharing electrons

When halogen atoms cannot get enough electrons to make halide ions they share electrons and form molecules. Two halogen atoms share outer electrons to form a halogen molecule. They also form covalent compounds with other non-metallic elements.

Hydrogen fluoride, hydrogen chloride, hydrogen bromide and hydrogen iodide are collectively known as the **hydrogen halides**. They are colourless gases that dissolve in water to form acidic solutions. For example, hydrogen chloride gas dissolves in water to form hydrochloric acid.

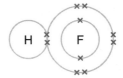

▲ Hydrogen fluoride.

Question

d *The diagram shows the electronic structure of a hydrogen fluoride molecule.*
(i) Give the formulae of (a) hydrogen chloride, (b) hydrogen bromide, and (c) hydrogen iodide.
(ii) Draw a diagram to show the electronic structure of hydrogen chloride.

Many covalent compounds of carbon, hydrogen and a halogen have industrial, medical or environmental importance. For example trichloromethane, $CHCl_3$, was used for many years as an anaesthetic (chloroform) and is still used as a solvent, including as dry-cleaning fluid.

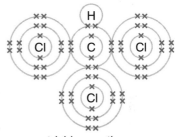

trichloromethane

Chlorofluorocarbons (CFCs), such as trichlorofluoromethane, were introduced as 'safe' alternatives to refrigerants like ammonia and sulfur dioxide. They were also widely used in aerosols. In the mid-1970s scientists identified CFCs as causing a 'hole' in the ozone layer of the atmosphere. This allowed too much UV radiation to reach the ground, causing increased risk of cataracts and skin cancer. In 1987 an international treaty, the Montreal Protocol, was signed. This called for reducing CFC use by 50% by 2000.

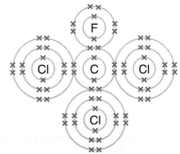

trichlorofluoromethane

Key points

- Halogens react with metals, making salts, ionic compounds that contain negative halide ions.
- As with metals, a more reactive halogen will displace a less reactive one in a compound.
- Halogens can also react with non-metals, like hydrogen, to make covalent compounds. The hydrogen halide gases are colourless and dissolve in water to form acidic solutions.

The transition elements

Role model metals

When we think of metals we think of hard, strong, shiny materials: metals like gold, copper and iron. These metals are found in the central portion of the Periodic Table, in a section called the transition elements. These elements, all metals, form a block between Groups 2 and 3.

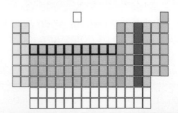

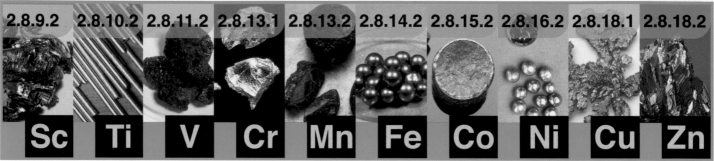

| 2.8.9.2 | 2.8.10.2 | 2.8.11.2 | 2.8.13.1 | 2.8.13.2 | 2.8.14.2 | 2.8.15.2 | 2.8.16.2 | 2.8.18.1 | 2.8.18.2 |
| Sc | Ti | V | Cr | Mn | Fe | Co | Ni | Cu | Zn |

▲ The first row of transition elements.

Transition elements are much harder than Group 1 metals; they do not scratch so easily.

Group 1 metals melt easily. Look at the graph. You can see that the melting points of transition elements vary, but they are all well above those for Group 1 metals.

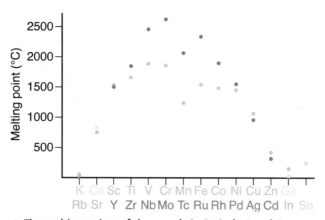

▲ The melting points of the metals in Periods 4 and 5.

Question

a Which have higher melting points, the transition elements from Period 4 or the transition elements from Period 5? Back up your answer using evidence from the graph.

Transition elements are also much less reactive than Group 1 metals. Look at the reactivity series, which is based on how easily the elements react with oxygen in the air and with water. The transition elements have middle to low reactivity, while the reactivity of Group 1 elements is notoriously high.

	Does it burn when heated in air?	Does it tarnish or corrode?	Does it react with cold water alone?
K	yes	yes	yes
Na	yes	yes	yes
Ca	yes	yes	yes
Mg	yes	yes	yes
Al	yes*	no	yes*
C	—	—	—
Zn	yes	yes	no
Cr	yes*	no	no
Fe	yes	yes	no
Ni	yes	yes	no
Sn	yes	yes	no
Pb	yes	yes	no
H	—	—	—
Cu	yes	yes	no
Hg	no	yes	no
Ag	no	yes	no
Pt	no	no	no
Au	no	no	no

most reactive

least reactive

* only if protective oxide layer is removed

▲ Reactions of metals in the reactivity series.

Much of a muchness

Elements in neighbouring groups of the Periodic Table vary enormously. Think about moving from Group 7 to Group 0. Fluorine is the most reactive element but the element next to it, neon, does not react at all. The transition elements in different columns are more similar. Adding an extra electron has much less effect. This is because you are not adding an *outer* electron. The electronic structures of the transition elements of Period 4 are shown at the top of the opposite page. As you move across you are adding electrons to the empty places in the third energy level (see page 167).

> **Question**
>
> **b** How many electrons would you expect silver, Ag, to have in its outermost energy level? Give your reasons.

Slow but pretty

Gold and platinum are renowned for their lack of reactivity, but copper and iron corrode and zinc is lively in comparison, reacting quickly with dilute acids.

> **Question**
>
> **c** Look at the reactivity series. Which transition element(s) would you expect to
> (i) occur as the metal in the environment?
> (ii) be isolated by reducing their compounds with carbon?
> (iii) react with dilute acids, making hydrogen?
> (iv) form an oxide layer that resists corrosion?

When transition elements react they often produce coloured compounds. Coloured glass, artists' pigments and some gems are beautiful colours because of the transition metal compounds they contain.

Once you start naming the compounds of transition elements, you meet numbers: copper II sulfate, iron III chloride. This is because many transition elements form more than one type of ion. Iron forms +2 ions and +3 ions. Copper forms +1 ions and +2 ions.

Look at the electronic structure of iron: 2.8.14.2. If it loses two electrons it will be $[2.8.14]^{2+}$ but if it loses three electrons it will be $[2.8.13]^{3+}$. It turns out that these two electron arrangements are both stable enough to exist, so both are formed.

Crucial catalysts

Many transition elements and their compounds act as **catalysts**. Platinum, rhodium and palladium are used in catalytic converters to clean up car exhaust fumes. Platinum is used as a catalyst in the production of nitric acid, iron when making ammonia and vanadium oxide during the production of sulfuric acid.

a

b

c

▲ (a) This 'ruby glass' is red because of gold compounds. (b) The emeralds are green because of chromium and iron compounds. (c) This pigment, 'cobalt blue', is a compound of cobalt.

> **Key points**
>
> - The transition metals are much less reactive, are stronger, harder, and have higher densities than the elements in Group 1.
> - As the atomic number increases along a row of transition elements, an inner energy level is being filled with electrons.
> - Many transition metals and their compounds make excellent catalysts.
> - Transition metal compounds are often coloured.

Unending progression?

Scientists have been discovering elements as elements for over 400 years. Before that, materials were discovered and used that later turned out to be elements. Look at the list in the margins. The elements with a question mark under the year have been used since ancient times.

Will we ever discover all the elements? Certainly the speed of discovery has slowed down. Now scientists are chasing very unstable elements, whose nuclei fall apart in tiny fractions of a second. Atoms of these elements are observed only in particle accelerators. These are huge machines that use magnetic fields to accelerate subatomic particles like neutrons and electrons. These are then crashed into atoms of known elements in the hope that the atom's nucleus will break down and make new atoms.

◄ This particle accelerator is 27 km around the circumference.

Just as the inside of a particle accelerator is an extreme environment, there are other extreme environments out in space. Think about the interior of a star, or the edge of a black hole, or the surface of a neutron star. Maybe the scientists of the future will find new elements in these places, stabilised by the incredibly extreme conditions.

◄ Scientists think there is a black hole at the centre of the Milky Way.

Element		Year
Antimony	Sb	?
Arsenic	As	?
Carbon	C	?
Copper	Cu	?
Gold	Au	?
Iron	Fe	?
Lead	Pb	?
Mercury	Hg	?
Silver	Ag	?
Sulfur	S	?
Tin	Sn	?
Zinc	Zn	?
Phosphorus	P	1669
Platinum	Pt	1735
Cobalt	Co	1739
Nickel	Ni	1751
Bismuth	Bi	1753
Hydrogen	H	1766
Nitrogen	N	1772
Chlorine	Cl	1774
Manganese	Mn	1774
Oxygen	O	1774
Molybdenum	Mo	1778
Tellurium	Te	1782
Tungsten	W	1783
Uranium	U	1789
Yttrium	Y	1789
Zirconium	Zr	1789
Strontium	Sr	1790
Titanium	Ti	1791
Chromium	Cr	1797
Beryllium	Be	1798
Niobium	Nb	1801
Vanadium	V	1801
Tantalum	Ta	1802
Cerium	Ce	1803
Iridium	Ir	1803
Osmium	Os	1803
Palladium	Pd	1803
Rhodium	Rh	1803
Potassium	K	1807
Sodium	Na	1807
Barium	Ba	1808
Boron	B	1808
Calcium	Ca	1808
Magnesium	Mg	1808
Iodine	I	1811
Cadmium	Cd	1817
Lithium	Li	1817
Selenium	Se	1817
Silicon	Si	1824
Aluminium	Al	1825
Bromine	Br	1826
Thorium	Th	1828
Lanthanum	La	1839
Erbium	Er	1843
Terbium	Tb	1843
Ruthenium	Ru	1844

Questions

a In the list, platinum was discovered in 1735. We attribute this discovery to a scientist called Antonio de Ulloa. We now know that, from ancient times, the natives of South America knew about, and used, platinum. Should Antonio de Ulloa be credited with the discovery, or should platinum be put with the other elements like gold and silver which have been known since ancient times?

b (i) Plot a bar chart to show the pattern of discovery of the elements. Put them into 50-year groups, e.g. 1600–1649, 1650–1699, with the 'ancient' elements as one group at the start.
(ii) What 50-year period was the 'best' for discovering elements?
(iii) Suggest why the rate of discovery has slowed down.

Naming elements

There is a tradition that the scientist who discovers an element gets to name it. Interestingly, there have been fashions in naming elements. Look at the list of names. You may be able to pick out some of the trends.

Question

c Pick out and name two elements that were named for:
   (i) planets   (ii) countries   (iii) scientists.

We now have very strict rules for naming 'new' elements. These apply to all elements with atomic numbers greater than 103. They were set out in 1978.

It was assumed that there would be many of these elements. So that scientists could discuss the undiscovered elements, they were given names based on their atomic numbers. 'Un-un-unium' is element 111, and 'ununbium' is element 112.

This 'temporary' name is used until the element has been discovered and the discovery has been confirmed by separate groups of scientists. This can take some years. Ununbium has been observed (at 22:37 on 9 February 1996) and the observation published. Until more groups of scientists confirm the discovery, the element will continue to be called ununbium. When the discovery is confirmed the group of scientists who made the first observation get to name the element.

Not quite there

In section 1.1, you read about an element called didymium. This was discovered by Carl Mosander in 1841, only to be found to be a mixture of two other elements in 1885. In the meantime, it caused scientists like Mendeleev problems because they tried to make the 'element' fit into their Periodic Tables. Similar problems happen today. If you look at a Periodic Table from 1999 you may see the element ununoctium included. This discovery was announced, but the scientists retracted their claim when they were unable to reproduce their results.

Questions

d What is the atomic number of the theoretical element called unbiunium?
e What do you think of the procedure for naming new elements? Is it fair on the scientists who make the discovery?
f What do you think of the scientists who announced the discovery of ununoctium and then retracted?

Key points

- Scientists have been seeking out, identifying, naming and characterising elements for the last 450 years. There have been fashions in the way scientists name elements, but there is now an agreed system.
- The search for new elements continues as scientists create extreme conditions, or observe them out in space.
- Observations of new elements generate excitement among scientists. If no-one else can find them, though, eventually the discoverers have to admit they made a mistake.

Element		Year
Samarium	Sm	1853
Caesium	Cs	1860
Rubidium	Rb	1861
Thallium	Tl	1861
Indium	In	1863
Helium	He	1868
Gallium	Ga	1875
Ytterbium	Yb	1878
Holmium	Ho	1879
Scandium	Sc	1879
Thulium	Tm	1879
Gadolinium	Gd	1880
Neodymium	Nd	1885
Praseodymium	Pr	1885
Dysprosium	Dy	1886
Fluorine	F	1886
Germanium	Ge	1886
Argon	Ar	1894
Europium	Eu	1896
Krypton	Kr	1898
Neon	Ne	1898
Polonium	Po	1898
Radium	Ra	1898
Xenon	Xe	1898
Actinium	Ac	1899
Radon	Rn	1900
Lutetium	Lu	1907
Protactinium	Pa	1913
Hafnium	Hf	1923
Rhenium	Re	1925
Technetium	Tc	1937
Francium	Fr	1939
Astatine	At	1940
Neptunium	Np	1940
Plutonium	Pu	1941
Americium	Am	1944
Curium	Cm	1944
Promethium	Pm	1944
Berkelium	Bk	1949
Californium	Cf	1950
Einsteinium	Es	1952
Fermium	Fm	1952
Mendelevium	Md	1955
Nobelium	No	1958
Lawrencium	Lr	1961
Rutherfordium	Rf	1964
Dubnium	Db	1967
Seaborgium	Sg	1974
Bohrium	Bh	1976
Meitnerium	Mt	1982
Hassium	Hs	1984
Darmstadtium	Ds	1994
Roentgenium	Rg	1994
Ununbium	Uub	1996
Ununquadium	Uuq	1998
Ununhexium	Uuh	2001
Ununpentium	Uup	2004
Ununtrium	Uut	2004

Humans have always lived with acids and alkalis. The acidity of the soil influences the plants that grow and the animals that feed on them. Soap, which is made using alkali, has been found in clay cylinders thought to be from 2800 BC. Vinegar, an acid, is thought to be the world's oldest cooking ingredient, going back 10 000 years.

Acidity around us

Left alone, the majority of the British landscape, where the soil is mildly acidic, would end up as oak woodland. Chalky areas, where the soil is slightly alkaline, would end up as beech woodland. Of course most of the countryside is farmed, and farmers manipulate the acidity of the soil according to the needs of their crops.

▲ Oak woodland (left) and beech woodland (right).

The acidity of rainwater varies depending on the air it has fallen through. Rainwater is usually slightly acidic, but if it has a pH of less than 5.6 it is considered to be **acid rain**. Pollutants in the air, oxides of sulfur and nitrogen, dissolve in the rainwater, making it acidic. Acid rain can kill trees and disrupt the pH of lakes and streams. The result is that aluminium is freed from compounds in the rocks and sediments, making fish produce too much mucus on their gills, smothering them, or disrupting enzymes needed for fish eggs to hatch.

▲ Blue and pink hydrangeas.

▲ Artist's impression of Ibn Zakariya al-Razi.

Indicators

Flowers are a natural indicator of the soil's acidity: hydrangeas are blue when grown on soil with a pH of 5.5 or lower, but pink when grown on alkaline soils where the underlying rock is chalk. The first indicator of acidity routinely used by scientists was **litmus**, which was made by the Dutch from lichen using a secret process from the sixteenth century onwards, a monopoly broken by an English producer only in 1940. Nowadays scientists have a wide assortment of indicators that are sensitive to pH. The most widely used of these is **universal indicator**, a mixture of three indicators that work together to produce a range of colours depending on the pH of the solution.

Useful acids

Acids are used in a wide range of manufacturing processes, from pickling food to making fertilisers, dyes, pigments and plastics. Sulfuric acid, one of the most commonly used chemicals in industry today, was first discovered in the ninth century by the physician, philosopher and alchemist Ibn Zakariya al-Razi, who was born and worked in the part of the world we now call Iran.

Back to bases

As you will learn in this section, you cannot have an acid without a **base**. A base is a substance that reacts with an acid, reducing its acidity. Soluble bases are called **alkalis**. The word alkali comes from an Arabic word meaning 'roasting'. This comes from the soap-making process. Ashes were heated with slaked lime (calcium hydroxide) to make the substances we now know as sodium hydroxide and potassium hydroxide. These 'alkalis' were then used to make soap.

Think about what you will find out in this section

How our understanding of acids and bases developed.	The relative contribution of different scientists to acid–base theory.
Modern definitions of 'acids' and 'bases'.	The meaning of strong and weak acids or alkalis.
How to 'titrate' acids and alkalis.	

What it does

Acids were first identified by their taste: 'acid' comes from the Greek word meaning sour. Look at the ingredients of the sweets. It is the acetic acid (also known as ethanoic acid) and the malic acid that make them sour. By the seventeenth century acids were identified by their behaviour: Robert Boyle, a British scientist, defined acids as substances that had a sour taste, were corrosive, changed litmus from blue to red, and became less acidic when combined with alkalis. You would probably get the same answer out of most young science students today.

What it contains

Simple definitions based on behaviour were not enough for later scientists. They wanted a definition that was based on what acids contained.

In 1838, Justus von Liebig suggested that acids contained one or more hydrogen atoms. But he realised that this was not enough to make a substance an acid: too many substances were known that contained hydrogen atoms but were not acids. Liebig's definition was that acids contained one or more hydrogen atoms *which could be replaced by metal atoms to form salts*. His definition was a combination of what the acid *contained* and what it *did*.

> ### Question
>
> **a** Why did Liebig need two parts to his definition of an acid?

Too big a jump

Science usually develops in little steps. Scientists share their evidence and their ideas. The theory moves on gradually. You often have many scientists making similar proposals at the same time, like Mendeleev and Meyer (see section 1.1), but once in a while someone suggests something really different, based on a completely new way of thinking. Once such scientist was Svante Arrhenius.

In 1887, Arrhenius proposed that acids were substances that produced hydrogen ions when put into water. To us this does not sound radical or shocking, but to the scientists of the day it seemed ridiculous. They firmly believed that ions could not exist in solution. Ions, particles with a charge, were associated with electricity, not chemistry. They had already rejected his idea that salts formed ions when they dissolved, which he had made three years earlier. He was considered a young fool, and his professors gave him only a fourth-class degree.

GB Fruit Flavour Pastilles with a Sour Sugar Coating Ingredients: Glucose syrup, Sugar, Modified maize starch, Gelatine, Acids (Malic, Acetic), Hydrogenated vegetable oil, Flavourings, Acidity regulator (E333), Colours (E104, E129, E122, E142). CONTAINS SULPHITES. MAY CONTAIN TRACES OF MILK AND WHEAT. **F** Pastilles Aux Arômes De Fruits Enrobées De Sucre

▲ The acids named in the ingredients give the sweets a sour taste.

Acids
H_2SO_4
HNO_3
HCl
H_3PO_4
H_2CO_3
HBr
HI
HF
CH_3CO_2H

Luckily for Arrhenius (and the future of chemistry!) some of his fellow scientists recognised the quality of his work and recommended him for jobs in science. Slowly, chemists came to see that he was right. His way of thinking helped them to understand the links between electricity, electrolysis, dissolving and acids. In the 1890s other scientists discovered protons and electrons, and suddenly Arrhenius' ideas about acids were mainstream. He was awarded the Nobel Prize for Chemistry in 1903.

Taking it further

The next big step in understanding acids came in 1923. Two scientists working independently, Johannes Brønsted and Thomas Lowry, proposed that acids were 'hydrogen ion donors' and bases were 'hydrogen ion acceptors'. They suggested that any acid–base reaction involved the movement of a hydrogen ion from one substance to another.

Unlike Arrhenius' earlier insight, this was very much an idea that fitted into the thinking of its time. Proposed by two scientists in the same year, it was swiftly accepted, and the terminology developed in line with it. As a hydrogen ion is the same as a proton, acids became **proton donors** and bases became **proton acceptors**.

Brønsted and Lowry also pointed out that when you put an acid in water an acid–base reaction takes place. The water acts as a base, accepting the protons. Each proton attaches to a water molecule, forming a positive ion of H_3O^+. For simplicity, we will still call them 'hydrogen ions' or 'hydrated hydrogen ions' and write them as H^+ or $H^+(aq)$, but they are really H_3O^+ or **oxonium ions**.

Arrhenius' ideas about acids and alkalis

Acids are substances that form hydrogen ions when put into water. Another ion is also formed.

$$HCl \rightarrow H^+ + Cl^-$$

$$H_2SO_4 \rightarrow 2H^+ + SO_4^{2-}$$

Alkalis are substances that form hydroxide ions when put into water.

$$NaOH \rightarrow Na^+ + OH^-$$

$$NH_3 + H_2O \rightarrow NH_4^+ + OH^-$$

Brønsted–Lowry ideas about acids and bases

Acids are substances that give protons to a base.

Bases are substances that accept protons from an acid.

$$\begin{array}{ccc} HCl & + & NaOH & \rightarrow NaCl + HOH \\ \text{proton donor} & & \text{proton acceptor} \end{array}$$

Questions

b What are (i) the similarities and (ii) the differences between Arrhenius' ideas about acids and the Brønsted–Lowry ideas about acids?

c Who do you think made the biggest contribution to our understanding of acids today? Give your reasons.

d Write two different balanced equations for hydrogen chloride gas reacting with water.

Key points

- Science often progresses in little steps, with many scientists moving forward with similar ideas.
- Sometimes a completely new approach is needed. What looks at first like a crazy idea turns out to be right.
- Acids are substances that can donate protons. Bases accept protons.

Eat it or avoid it?

We put vinegar, a solution of ethanoic acid, on our food. When we handle a similar concentration of hydrochloric acid, $0.8\,mol/dm^3$, we wear safety glasses and take care not to get it on our skin. Why the difference?

The formulae do not tell us the answer. Hydrogen chloride contains one acidic hydrogen, HCl, and so does ethanoic acid, CH_3CO_2H (the other three hydrogens in ethanoic acid are not involved in acid–base reactions). Both substances could make the same number of hydrogen ions when put in water.

But they don't. Hydrochloric acid forms the maximum number of hydrogen ions, but ethanoic acid does not. All the ethanoic acid dissolves, but it does not all **ionise**. About 95% of the ethanoic acid stays as ethanoic acid molecules and only about 5% ionises to make hydrogen ions. We see this when we measure the pH: $0.8\,mol/dm^3$ of hydrochloric acid produces enough hydrogen ions to have a pH of 0.1, but $0.8\,mol/dm^3$ of ethanoic acid produces only enough hydrogen ions to have a pH of 2.4.

$$HCl(g) \xrightarrow{H_2O} HCl(aq) \rightarrow H^+(aq) + Cl^-(aq)$$
$$\quad\quad\quad\quad\quad 0\% \quad\quad\quad\quad 100\%$$

$$CH_3CO_2H(l) \xrightarrow{H_2O} CH_3CO_2H(aq) \rightleftharpoons H^+(aq) + CH_3CO_2^-(aq)$$
$$\quad\quad\quad\quad\quad\quad 95\% \quad\quad\quad\quad\quad 5\%$$

▲ We eat citric acid in citrus fruit.

Question

a Write balanced equations of (i) sulfuric acid and (ii) nitric acid fully ionising in water.

Hydrochloric acid is a **strong acid** and ethanoic acid is a **weak acid**. Strong acids fully ionise in water. Weak acids do not. Lemons seem sour enough to the taste, but citric acid does not ionise fully. Nitric acid, however, does – it is a corrosive and dangerous chemical. Some strong and weak acids are shown in the table.

Understanding pH

Scientists needed a measure of acidity that was easy to use. This was difficult because strong acids produce high concentrations of hydrogen ions, while weak acids produce much lower concentrations. What about pure water? It contains about $0.0000001\,mol/dm^3$ hydrogen ions. Then there were alkalis to consider!

Name	Formula
Strong acids	
Hydrochloric acid	HCl
Sulfuric acid	H_2SO_4
Nitric acid	HNO_3
Weak acids	
Ethanoic acid	CH_3CO_2H
Citric acid	$CH_2(CO_2H)COH(CO_2H)CH_2(CO_2H)$
Carbonic acid	H_2CO_3

They solved the problem by creating the 'pH scale', where moving from pH 2 to pH 1, for example, increases the hydrogen ion concentration 10-fold. This approach is often used by scientists when they are trying to express a huge range of possible values on a simple scale. Another example of a scale using powers of 10 is the decibel scale for sound.

pH	Concentration of $H^+(aq)$ in mol/dm^3
−2	100
−1	10
0	1
1	0.1
2	0.01
3	0.001
4	0.0001
5	0.00001
6	0.000001
7	0.0000001
8	0.00000001
9	0.000000001
10	0.0000000001
11	0.00000000001
12	0.000000000001
13	0.0000000000001
14	0.00000000000001

Question

b What concentration of hydrogen ions corresponds to pH 7?

Alkalis

There are strong alkalis and weak alkalis, like strong acids and weak acids. A strong alkali is one that fully ionises in water, producing many hydroxide ions, $OH^-(aq)$. Examples of strong alkalis include potassium hydroxide and sodium hydroxide, which fully ionise in solution. A weak alkali produces fewer hydroxide ions when put in water. An example of a weak alkali is ammonium solution.

$$NaOH(aq) \rightarrow Na^+(aq) + OH^-(aq)$$
$$\quad 0\% \qquad\qquad 100\%$$

$$NH_3(aq) + H_2O(l) \rightleftharpoons NH_4^+(aq) + OH^-(aq)$$
$$\quad 99.6\% \qquad\qquad\qquad 0.4\%$$

Question

c Write a balanced equation to show what happens when you put potassium hydroxide in water.

No limits

So far we have been thinking about the acids we meet in our everyday lives or in school science laboratories. The Arrhenius definition of acids works very well for these. But scientists have gone beyond such everyday situations, looking for stronger and stronger acids. These 'super-acids', first discovered in the 1930s and millions of times more acidic than everyday strong acids, are better explained using the Brønsted–Lowry theory of acids. They force hydrogen ions (protons) onto even the most reluctant recipient. In the 1970s scientists went even further, creating 'magic acids' that are thousands of trillions times stronger than 1 mol/dm^3 H$_2$SO$_4$.

Question

d Explain why Brønsted–Lowry acid–base reactions can take place without water.

Super-strong bases were discovered much earlier, back in the 1850s. These can 'rip' hydrogen ions off even the most reluctant donor. They are used widely in the chemical industry to make stable substances into unstable substances, which will then react in the way the scientist wants.

▲ George Olah, who received a Nobel prize in 1994 for 'magic acids'.

Key points

- Normally, substances behave as acids and alkalis only if water is present. Acids produce hydrated protons, while alkalis produce hydroxide ions.
- Weak acids or alkalis do not ionise completely in water.
- Scientists' ideas about acids and bases are still evolving.

Neutralise!

We often need to neutralise acids. Farmers may want to alter the pH of their soil. Water companies have to make sure that the water we drink is close to neutral. Environmental scientists may need to balance the effect of acid rain in lakes, or 'scrub' acidic gases from factory smoke. Chemical engineers have to put exactly the correct amount of acid with the correct amount of base to make the perfect salt.

Measure it

To carry out these tasks successfully, we need to be able to measure the concentration of acid present. We do this by carrying out a **titration**. The photo shows traditional titration equipment. The long, thin tube with a tap at the bottom is called a **burette**. This is a very precise instrument for measuring volume; the smallest divisions are $0.1\,cm^3$. A burette takes $50\,cm^3$ of liquid.

▲ Titration equipment.

Question

a A lab technician knows he can be $0.05\,cm^3$ out on any burette reading.
(i) What percentage is this of (a) $10\,cm^3$ and (b) $40\,cm^3$?
(ii) Why does the lab technician say it is more accurate to use the burette to measure larger volumes than smaller volumes?

You put the solution with the unknown concentration of acid in the conical flask. It is very important to measure the volume of this solution precisely. You can do this using a **pipette**, like the ones in the photo.

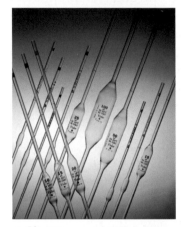
▲ Pipettes.

Question

b (i) In the old days people used to suck the liquids into the pipette using their mouths. Now they use a pipette filler. Why?
(ii) The middle part of the pipette is wide, so it can hold enough liquid, but the top section, with the measuring line, is very narrow. Why?

You fill the burette with an alkaline solution with a known concentration. The solution in the burette in the photo is $0.1\,mol/dm^3$ sodium hydroxide solution. You must make sure that there are no bubbles or air spaces in the burette, including below the tap. You then read the burette. You make sure your eye is level with the top of the liquid and always read the bottom of the curved surface, called the **meniscus**. Remember, when the burette is full you have used no solution, so the zero is at the top and the reading increases as you move down and use more liquid.

▲ Reading a burette.

You then add a few drops of indicator to the solution in the conical flask. In the photo of the titration equipment you can see that the indicator is pinky-red in the acidic solution in the conical flask.

Question

c Why must your eye be on the same level as the top of the liquid?

You then open the tap and run in the alkaline solution. It is important to make sure that you mix the solution from the burette with the solution in the conical flask. Experienced hands can swirl the conical flask with one hand and control the burette tap with the other!

The idea is to stop adding the alkali as the indicator *just* changes colour, when the solution is *exactly* neutral. This is nearly impossible to do the first time, so the first attempt is always a 'rough', which gives you an idea of the correct volume of alkali solution to add. You then repeat the process, slowing the addition of the alkali solution to drips as you approach the volume necessary to exactly neutralise the acid. When the indicator is *just* changing colour you stop adding the alkaline solution. This is the **end-point**. You then read the burette and work out how much alkaline solution you have used. This volume is your first **titre**.

You then repeat the process until you get two titres within $0.1\,cm^3$ of each other. This is done for two reasons. Firstly, you can check that the technique is reproducible: if you cannot get two results within $0.1\,cm^3$ of each other then something is wrong. Secondly, you can then take an average of the two titres and get as close as possible to the correct volume, so improving the accuracy of the measurement. If you are careful, and there are no unanticipated problems, you should have to do only two or three titrations.

A result

The table shows the results of an investigation to find out the concentration of hydrochloric acid in a solution. There was $0.1\,mol/dm^3$ sodium hydroxide solution in the burette, and $25\,cm^3$ of the acidic solution in each conical flask. You have to choose the right indicator – in this case methyl orange indicator was used.

Question

d Why is it important to mix the contents of the conical flask?

Question

e Look at the photo of the titration equipment. Why is the conical flask on a white surface?

Burette reading 1	Burette reading 2	Titre (cm^3)
0.0	21.2	21.2
21.2	38.6	17.4
0.0	17.5	17.5

Questions

f (i) What is the 'rough titre'?
   (ii) What are the two more accurate titres?
   (iii) Work out the average (mean) of the accurate titres.
   (iv) Why wasn't a third accurate titration done?
g Why was the burette refilled after the first accurate titration?

Key points

- Titration is the method used to measure the concentration of a substance in solution.
- Precision is important: you need to use equipment that can measure volumes to $0.1\,cm^3$.
- Careful observation is necessary. You watch for the end-point, where the indicator just changes colour.
- Reproducibility is important. You should repeat the titration until you get two very close results, then take the mean (average) to improve the accuracy of your results.

Recap

On the previous page you read about an investigation to find the concentration of acid in a solution. $17.45\,cm^3$ of $0.1\,mol/dm^3$ sodium hydroxide solution was needed to neutralise $25\,cm^3$ of the acidic solution.

Working it out

$$NaOH(aq) + H^+(aq) \rightarrow Na^+(aq) + H_2O(l)$$

This equation shows us that 1 mole of sodium hydroxide is needed to neutralise 1 mole of hydrogen ions. This means that if we calculate the number of moles in the titre, we will know the number of moles of hydrogen ions in $25\,cm^3$. You also need to remember that $1\,dm^3$ is $1000\,cm^3$.

> $1000\,cm^3$ of $0.1\,mol/dm^3$ $NaOH(aq)$ contains 0.1 moles of NaOH.
>
> $1\,cm^3$ of $0.1\,mol/dm^3$ $NaOH(aq)$ contains 0.0001 moles of NaOH.
>
> $17.45\,cm^3$ of $0.1\,mol/dm^3$ $NaOH(aq)$ contains $17.45 \times 0.0001 = 0.001745$ moles NaOH.
>
> So there are 0.001745 moles of $H^+(aq)$ in $25\,cm^3$ of the acidic solution.
>
> There will be 40 times as much in $1000\,cm^3$ of acidic solution, which is 0.0698 moles.

So the concentration of hydrogen ions in the acidic solution is $0.07\,mol/dm^3$.

Why methyl orange?

Choosing the correct indicator is important. We can check this by monitoring the pH changes during the titration. Look at the three graphs. They show the change of pH during three different titrations. The pH readings were taken during the titrations, as in the photo.

Question

a The investigation was repeated for a different acidic solution. This time $28.2\,cm^3$ of $0.1\,mol/dm^3$ $NaOH(aq)$ was needed to neutralise $25\,cm^3$ of the acidic solution.
(i) Is this sample more or less acidic than the first?
(ii) Work out the concentration of hydrogen ions in this second solution.

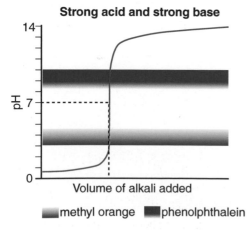

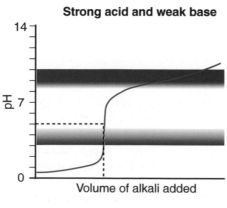

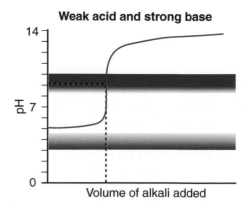

methyl orange phenolphthalein

Also marked on the graphs are two indicators. Methyl orange is red in more acidic conditions and yellow in more alkaline conditions. It changes colour at about pH 3–4. Phenolphthalein is colourless in more acidic solutions and pink in more alkaline conditions. It changes colour at about pH 9–10.

Look at the left-hand graph. It shows what happens when you titrate a strong acid, like hydrochloric acid, with a strong alkali, like sodium hydroxide. The jump in pH at neutralisation is very big, from pH 1 to pH 12, and it happens across a very small volume of alkali, so the graph goes vertically up. As you can see, both methyl orange and phenolphthalein change colour within the steep rise of pH. Both indicators would change colour within less than 0.1 cm³ of added alkaline solution, so both will give an accurate volume for the end-point.

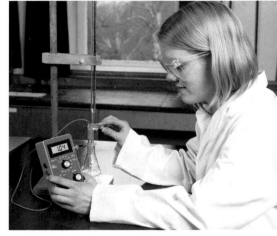
▲ Measuring pH during a titration.

Look at the middle graph. It shows what happens when you titrate a strong acid, like hydrochloric acid, with a weak base, like ammonia solution. This time the jump in pH happens between pH 1 and pH 7. The methyl orange changes colour during this jump, so it is a good indicator to use. The phenolphthalein changes colour at pH 9–10, which is after the acid has been neutralised.

The right-hand graph shows what happens when you titrate a weak acid, like ethanoic acid, with a strong base, like sodium hydroxide. This time the pH at neutralisation, shown by the jump in pH, is between pH 7 and pH 11. The methyl orange is not going to work. It will go yellow as soon as you put it in the weak acid. Phenolphthalein is the correct indicator to choose.

Questions

b Look at the graph in the margin. It shows the reaction between a weak acid and a weak alkali. Is methyl orange and/or phenolphthalein a suitable indicator? Give reasons for your answer.

c You are planning a titration between (i) citric acid and potassium hydroxide, and (ii) sulfuric acid and ammonia solution. Which indicator would you choose for each titration?

Strong acid and strong base

pH

14

7

0

Volume of alkali added

■ methyl orange ■ phenolphthalein

Key points

- Titration can be used to measure the concentration of acid in solution.
- Different indicators are suitable for different acid–base titrations, depending on the pH of the end-point.
- We record what volume of the alkali solution we need to neutralise a given volume of the acidic solution, then work out the number of moles of alkali we used.
- We then use the equation to work out how many moles of acid were present. Knowing the number of moles and the volume, we then work out the concentration of the acid.

▲ Routine titrations in modern labs are carried out using titration machines.

Planning

Mary has been given some tablets for indigestion from a health food shop. She thinks the active agent may be a base or bases that neutralise excess acid in the stomach. She decides to measure the amount of base by titration, using one tablet for each titration. She grinds up the tablet using a pestle and mortar and adds water. Some of it dissolves and some does not. She washes it all into a conical flask.

To carry out some preliminary experiments, Mary needs to choose an acid and an indicator. She does not know if the indigestion tablet contains strong alkali or weak alkali, or a combination of the two. She looks up indicators and finds the chart below.

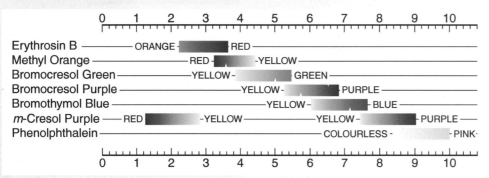

Mary chooses sulfuric acid and bromocresol green.

In her preliminary experiment, Mary fills her burette with 1 mol/dm³ sulfuric acid. She adds indicator to the partially dissolved tablet and starts to add the acid. The mixture in the conical flask starts to react, producing lots of froth that bubbles up the neck of the flask. The indicator changes colour before she has added 5 cm³. Mary's teacher suggests that the acid in the burette is too concentrated.

Questions

a   *Which acid would you choose? Give reasons for your choice.*

b   *(i) If the alkalis in the tablet are strong, which indicators could Mary use?*

(ii) If the alkalis in the table are weak, which indicators could Mary use?

(iii) Which indicators should Mary choose from?

Questions

c   *Describe the colour changes that would happen during the titration.*

d   *Give two separate reasons why reducing the concentration of the acid would be a sensible modification.*

Implementing

Mary settles for using 0.1 mol/dm³ sulfuric acid in the burette. She carries out her titrations carefully, adding the acid slowly and swirling the mixture after each addition, so she can cope with the frothing. She also notices that the powder in the mixture dissolves during the titration. Her results are shown in the table.

Titration	Titre (cm³)
Rough	34.6
1	32.6
2	32.3
3	32.4

Question

e   *Suggest why*
(i) the froth is produced, and
(ii) the tablet dissolves during the titration.

Analysing and concluding

Mary's teacher suggests that she uses the equation below for the neutralisation reaction, and that by using this equation, Mary should work out that each tablet contains 0.0065 moles of hydroxide ions.

$$H_2SO_4(aq) + 2OH^-(aq) \rightarrow 2H_2O(l) + SO_4{}^{2-}(aq)$$

Questions

f Mary decides to use 32.35 cm³ as her titre of sulfuric acid solution. Explain why.

g Show carefully, step by step, how Mary should work out that 0.0065 moles is the expected answer.

Mary thinks that the neutralising substance in the tablets is a carbonate. She decides to repeat the calculation, based on this equation:

$$H_2SO_4(aq) + CO_3{}^{2-}(aq) \rightarrow H_2O(l) + SO_4{}^{2-}(aq) + CO_2(g)$$

Question

h How many moles of carbonate ions would each tablet contain? Explain how you decided on your answer.

Evaluating

If all the neutralising substance is carbonate, then there is 0.003235 moles of carbonate ions in each tablet, which would produce 0.003235 moles of carbon dioxide. Mary researches the volume of gases and discovers that 1 mole of any gas occupies 24 dm³ at room temperature and pressure.

Question

i What volume of carbon dioxide would be made by reacting one tablet?

Mary carries out an experiment, reacting a tablet with acid and trapping the gas in a measuring cylinder. She measures 24 cm³ of gas. She decides that the tablet contains mostly carbonate compounds but some other bases or alkalis.

Questions

j What other explanations may there be for Mary's results?

k What do you think are the weaknesses and strengths of Mary's investigation?

Key points

- Trial experiments are important in investigations because the plan may need modification.
- When planning a titration, trial experiments combined with your knowledge will help you choose the correct concentration of reactant and the correct indicator.
- When you have drawn a conclusion, try to think of other experiments to test it.

Water

Three-quarters of our planet's surface is covered with water. Our bodies are about 60% water. It is the main ingredient of our blood and of the fluid that bathes our cells, each of which is 70–90% water. This means that water is fundamental to the way our planet functions, and to keeping humans alive and healthy. Understanding the chemistry of water is a necessary survival skill!

On tap

In Britain, we take water for granted. We turn on a tap and safe, drinkable water comes out. We have so much of it that we use it to wash our cars, water our gardens and flush our toilets. We even fill swimming pools with it! Occasionally, when rainfall has been lower than usual, we have what we call a 'water shortage'. We are asked not to use the hose. We may be asked to save water by taking showers rather than baths. We can put a brick in the toilet cistern so we waste less water in each flush.

Other countries are less fortunate. Drought is a killer. Even if there is enough water for the people, there may not be enough to produce food: crops do not grow and farm animals die, causing famine. If water is present, it may not be drinkable. Water that seems clean can contain dissolved toxins, or microscopic disease-causing organisms. Drinking seawater is lethal.

▶ A farmer waters dry fields in Guangzhou, China, in 2002.

Weird stuff

From a chemist's point of view, water is weird stuff. It is made up of small molecules, and substances with small molecules are usually gases at room temperature. Look at the graph. It shows the boiling points of substances made up of molecules with a relative molecular mass under 40. Water has a much higher boiling point than one would expect for its relative molecular mass of 18. This means that water is liquid across most of the Earth's surface.

Why is the boiling point so high? Because there are unusually strong intermolecular forces between water molecules, called **hydrogen bonds**. These give water other peculiar properties, too. For example, you need to put a lot of energy into water to heat it up: 4.2 J/g for every 1 °C rise. This means that bodies of water do not heat up or cool down quickly. This is true of seas, ponds and even organisms.

The hydrogen bonds also give ice its structure. To make hydrogen bonds the molecules have to line up in a certain way. Once ice forms, every molecule is in its place, each hydrogen atom bonded to four other molecules. To form this structure, the molecules have to be more spread out than those in liquid water. This makes ice at 0 °C less dense than water at 0–4 °C, so ice floats on water. This means that water freezes from the surface down rather than from the bottom up. As ice is also a very good insulator, the temperature has to be below 0 °C for a long time before a pond or lake will freeze all the way to the bottom. This gives living things the chance to survive in the Arctic, the Antarctic or just through a particularly harsh winter in Britain.

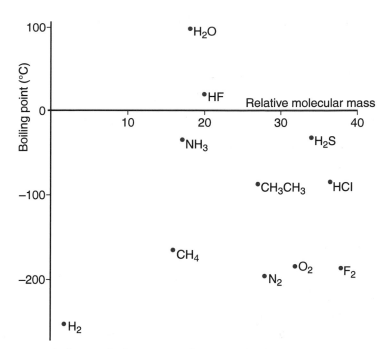

▲ Graph showing boiling points of materials made up of small molecules.

Think about what you will find out in this section

Where on Earth is the water?	What other substances do we find in the water around us?
How much of various substances can dissolve?	How do we make water drinkable?
Does hard water matter?	

Around and around

Scientists now know in detail where the water is on the Earth and how it moves. This helps us to monitor and predict the weather, and to anticipate how our climate may change over time. The diagram shows scientists' ideas about the water cycle. Water exists in various places and forms called reservoirs, and moves between these by various processes. For example, water is heated by the Sun and evaporates from the oceans, from surface water on land and from plants. When the air cools, the water vapour condenses, forming clouds of liquid water that will eventually fall as rain.

Question

a (i) Add up the mass of water in all the reservoirs.
(ii) What percentage of the water is moved each year by (a) evaporation, and (b) rivers?
(iii) What percentage of the stored water is in ice and snow?

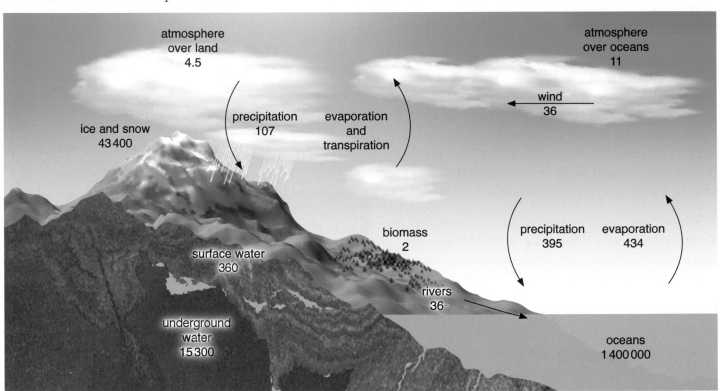

▲ The water cycle. Volumes of reservoirs in teratonnes (1 teratonne = 1 000 000 000 000 tonnes). Arrows show moving water, in teratonnes per year.

Acid clouds

As vapour, the water mixes with the other gases of the air. The amount of water vapour in the air varies, depending on where it is and the temperature. When the water condenses and forms clouds, other substances dissolve in the water droplets. Carbon dioxide is soluble in water. Water in the clouds is already a mixture: the dissolved carbon dioxide forms a weak acid, carbonic acid. This makes rainwater slightly acidic. Up to 0.14 g of carbon dioxide will dissolve in 100 g of water. We call this ratio the **solubility** of carbon dioxide in water. Sulfur dioxide is a much more soluble gas: up to 10.6 g of sulfur dioxide will dissolve in 100 g of water. If clouds and rain mix with air polluted with sulfur dioxide, the sulfur dioxide will react with the water, forming an acidic solution. This acid is stronger than carbonic acid, reducing the pH below 5.6 and producing acid rain.

Question

b Give two reasons why sulfur dioxide has more effect on the pH of the rain than carbon dioxide.

Other gases

Oxygen is not very soluble in water. That is why we need haemoglobin to carry oxygen in our blood. In fact, the solubility of oxygen in water at 25 °C and 1 atmosphere pressure is only 0.00082 g per 100 g. Even so, this is enough to keep fish and other aquatic organisms alive.

Like ammonia, sulfur dioxide and other gases, the solubility of oxygen decreases with temperature. Look at the table. At 40 °C there is only 44% of the oxygen that was present at 0 °C. At higher temperatures, the molecules of oxygen have more energy and are more likely to escape from the solution and become part of the gas above.

Temperature (°C)	Solubility of oxygen in water (g/100 g)
0	0.00146
10	0.00113
20	0.00091
30	0.00075
40	0.00064

Question

c Some factories and power stations release clean but hot water into rivers. Suggest how this may affect wildlife.

Another factor that affects the solubility of a gas is the pressure. The higher the pressure, the more gas dissolves. This is how we make fizzy drinks: carbon dioxide is forced into the solution under pressure. It is also why divers experience 'the bends'. The pressure is greater deep under the water, so more nitrogen from the air dissolves in the blood and tissue fluid. If the diver comes up too quickly, there will be a sudden decrease in pressure. This is like releasing the cap of a fizzy drink bottle. The bubbles of nitrogen can cause agonising pain and may kill if they interfere with blood flow.

Question

d Explain why using a decompression chamber decreases the danger of developing 'the bends'.

What dissolves?

Most gases are made of small, covalent molecules. Covalent substances can be soluble or insoluble in water: it depends on whether the molecules are attracted to water molecules or not. Giant molecular structures, like diamond or sand or plastics, are insoluble in water.

Key points

- The water cycle shows where Earth's water is and how it moves around.
- Liquid water in the air, in both clouds and rain, dissolves soluble substances in the air.
- The solubility of the substance indicates how much can dissolve in 100 g of solvent to form a saturated solution.
- Some covalent substances are soluble in water, but many are not, particularly those with very large molecules or giant structures.
- Gases are less soluble at higher temperatures.

Nor any drop to drink

Too salty

There are many soluble ionic compounds in the environment and these dissolve in water as it flows through the ground on its way to the sea. Look back at the water cycle diagram. The biggest reservoir of water is the oceans, so most of the water on the earth is seawater. Seawater contains many dissolved ionic compounds, the most abundant of which is sodium chloride. These dissolved compounds make it deadly to drink.

Seawater contains 3.5 g of salts in every 100 g of water. However, it is not saturated with salt. Look at the graph. It shows **solubility curves** for various ionic substances, including sodium chloride. Pick out the curve for sodium chloride. About 36 g will dissolve in 100 g of water at 25 °C, so seawater is only about 10% saturated.

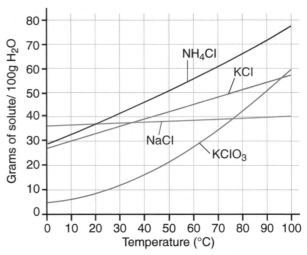
▲ Solubility curves.

Question

a (i) Which is the most soluble ionic substance out of the four shown in the graph?
(ii) Is there more than one answer to (i)?

Temperature counts

Sodium chloride is different from most salts in that its solubility does not vary much with temperature. This means that sodium chloride is likely to crystallise out of solution because the water has evaporated, raising the concentration of the salt, as has happened to create the salt flat shown in the photo. Look at the solubility curve for ammonium chloride, NH_4Cl. If you cooled a saturated solution of ammonium chloride from 70 °C to 10 °C, the mass of ammonium chloride you could dissolve would drop from about 61 g to about 33 g; 28 g of ammonium chloride would crystallise out of solution.

Questions

b You have a solution at 95 °C that is saturated with both potassium chloride and potassium chlorate ($KClO_3$). How much (i) potassium chloride and (ii) potassium chlorate would crystallise out of solution if you cooled the mixture to 10 °C?
c Compare how the solubility of (i) gases and (ii) ionic solids varies with temperature (look back to section 3.1).

▲ A salt flat in Bolivia.

Too toxic

Many of the substances that can dissolve in water are toxic. To avoid accidental poisoning, the testing of our drinking water is strictly regulated. You can see a list of the substances tested for in the table. Water companies also test for microorganisms (see below) and for taste, colour and cloudiness.

But no system is perfect. In 1988, in north Cornwall, someone opened the wrong tap and 20 tonnes of aluminium sulfate were put directly into the water supply. As a result, water that contained unacceptably high levels of aluminium ions was supplied to 20 000 people. Now, years later, some consumers are still convinced that it had a long-term effect on their health.

Often a toxin that is present well below acceptable levels can be concentrated by plants or animals taking in the water. It is then passed along the food chain in toxic amounts. This happened with mercury and shellfish in Minamata Bay, Japan, in the 1950s: before the link was recognised and broken, 1400 people died. Selenium, a mineral needed at low levels for good health, is concentrated to toxic levels by certain plants that grow in the western USA. Cattle eating the plants are poisoned. Hijiki, a seaweed once thought edible, has recently been found to contain unacceptable levels of arsenic.

> **Question**
>
> *d* *Explain why it may be more dangerous to eat the plant or animal rather than drink the water.*

Microbe killers

Water can also transmit diseases. Examples are known of diseases caused by viruses (hepatitis A and polio), bacteria (cholera and typhoid) and protozoa (cryptosporidiosis) that are transmitted by water. Care must be taken to separate human sewage from water sources used for drinking water. In Britain we treat sewage before disposing of it well away from water supplies, test water regularly for bacteria and treat our water with chlorine.

> **Question**
>
> *e* *Why is it important to separate water supplies and sewage?*

A human right?

In November 2002, the United Nations Committee on Economic, Social and Cultural Rights declared that access to enough clean water for personal and domestic uses is a fundamental human right of all people. Even so, the World Health Organisation (WHO) believes that 1 billion people still do not have access to a safe supply of drinking water.

Our drinking water is tested for ...
1,2-dichloroethane
aluminium
ammonium ions
antimony
arsenic
benzene
boron
bromate ions
cadmium
chloride ions
chromium
copper
cyanide ions
fluoride ions
hydrogen ions
iron
lead
manganese
mercury
nickel
nitrate ions
nitrite ions
pesticides (various)
polycyclic aromatic hydrocarbons
radioactive isotopes
selenium
sodium ions
sulfate ions
total organic compounds
trichloroethane
trichloroethene
trihalomethanes

Key points

- Many ionic compounds dissolve in water. This means that most of the water on the planet contains too high a concentration of dissolved substances to be drinkable.
- Soluble solids are more soluble at higher temperatures.
- Water contaminated with poisonous substances can kill. Plants and animals can concentrate toxins from water.
- Water can also transmit diseases from one human to another. Sewage must be carefully treated and separated from sources of drinking water.

Find a source

Drinkable water is free of poisons and disease-causing microorganisms. The first step towards clean, safe drinking water is to find and test a suitable source. In Britain the source is usually rainwater, which is collected from groundwater or rivers and often stored in reservoirs. In drier countries it is often water deep underground, which sometimes has to be drilled for. Occasionally it is even seawater. This would be deadly in its natural state, so it is treated to remove the salt.

▲ A water reservoir in Wales.

Filter and refilter

The diagram below shows how water will be processed at a new water treatment works in East Ham, London. The water source is water trapped in chalk rocks underground. This is an excellent source, since chalk is a natural filter, so the water is already of good quality. The water is reached using boreholes and pumped to the water treatment plant. There it is processed through various filters called **filter beds** to remove small bits of solid. In the East Ham works these consist of rapid gravity filters to remove large bits, and granular activated carbon filters to remove microscopic bits and some dissolved substances (see below). The water is then treated with chlorine to kill microorganisms, including harmful bacteria and viruses.

▲ An artesian well in Morocco.

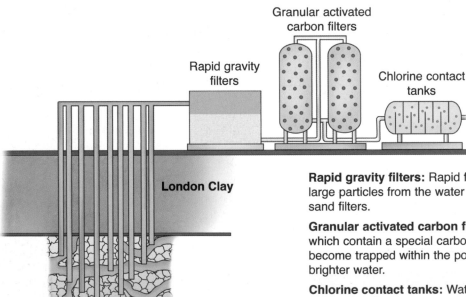

Rapid gravity filters: Rapid filtration is the physical removal of large particles from the water as it passes downwards through sand filters.

Granular activated carbon filters: Water is fed into vessels which contain a special carbon material. Microscopic particles become trapped within the pores of the carbon to leave a purer brighter water.

Chlorine contact tanks: Water is disinfected by dosing the water with chlorine to ensure the elimination of bacteria or viruses prior to distribution.

▲ Modern water treatment plant.

Question

a *Explain why the level of chlorine used to treat the water needs to be carefully controlled.*

More filters

The water that comes out of our taps is safe, but many of us are no longer content with the taste. There is a huge market for mineral water. Many offices have water coolers, supplied with bottled water. Some of us are installing extra water filters in our homes.

Some filters use activated carbon. This is charcoal which has been broken up into granules to increase the surface area and then treated to give the pieces a slight positive charge. Some contaminants in the water stick to the carbon, which has to be replaced on a regular basis.

Silver can also be used to purify water effectively, absorbing many impurities. The photo shows a low-tech but very effective silver filter. A sludge made up of tiny bits of silver is painted on the outside of unglazed ceramic pots, which are put inside a larger, impermeable container. Water is put inside the pot and seeps through the ceramic and past the silver. The clean water collects in the outer container.

Ion exchange is also used for water purification. The water is filtered though an insoluble, granular material called an **ion exchange resin**. This resin has a charged surface that attracts ions. Unwanted ions in the water are exchanged for preferred ions on the resin. You will learn more about how ion exchange resin works when you study water softening (section 3.5).

Stills

One way of purifying substances is to evaporate the water and allow it to recondense elsewhere, leaving the impurities behind. This process is called **distillation** and is carried out in a **still**. In a solar still the energy to evaporate the water is provided by the Sun.

Question

b Why is the carbon broken up to give a large surface area?

▲ 'Potters for Peace' silver water filters.

▲ A solar still.

Question

c Design your own solar still, suggesting materials to use for each part and including a diagram.

Key points

● Water is safe to drink only if the levels of dissolved salts, other toxic substances and microorganisms are kept below certain limits.
● Water intended for drinking must be drawn from appropriate sources, passed through filter beds and then sterilised.
● Water quality can be improved by filtering it through activated carbon, silver or an ion exchange resin.
● Distillation is another way of purifying water.

Scum and scale

In some parts of the country the inside of your kettle goes white and furry and your shower screen dries with white spots on it. These are both examples of **limescale**. In the same areas you will need to put more detergent into your washing and you get a scummy layer on the surface of water if you use soap. In these areas the water is **hard** water. In areas where the water is **soft**, lather forms readily so less soap is needed.

▲ Limescale in a kettle.

▲ Soap scum.

What makes water hard?

Salts dissolve in rainwater as it seeps through soil and rock. If those salts contain enough calcium and/or magnesium ions, then the water is hard. If the dissolved salts do not contain calcium and magnesium ions, the water is soft water. If one litre of the water has the same number of Ca^{2+} and/or Mg^{2+} ions as 200 mg of calcium carbonate, then the water is very hard.

Is it a problem?

When hard water is heated, insoluble calcium and magnesium compounds crystallise from the solution. These stick to the surface of the heating element in kettles or hot water tanks. The photo opposite shows limescale on a heating element. Scale like this makes heating less efficient. At home it makes washing machines, dishwaters, kettles and immersion heaters less energy-efficient and leads to higher energy

Soft to moderately soft

Slightly hard to moderately hard

Hard to very hard

Newcastle upon Tyne

Leeds

Manchester

Lincoln

Norwich

Birmingham

Cardiff

London

Bristol

Southampton

Exeter

Brighton

▲ Do you live in a hard water area?

© British Water

bills. The same is true for industrial heaters. The scale can also cause problems by decreasing water flow through industrial equipment.

Calcium and magnesium ions make soap and detergent less effective. The soap or detergent has to react with the calcium and/or magnesium ions before it can react with stains. In hard water areas you have to add more soap or detergent before getting lather.

The problem with soap is that it produces an insoluble scum when it reacts with the calcium and/or magnesium ions. This greyish, sticky precipitate makes washed clothes look dirty and sticks to baths or sinks. We have stopped using soap and started using detergents instead to avoid this problem. When detergents react with calcium or magnesium ions they produce soluble compounds. You still have to use more detergent than you would in a soft water area, but the clothes look clean and it is easier to clean baths and sinks.

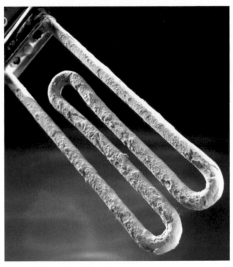
▲ Scale due to hard water.

Questions

a What problems are caused by having hard water (i) at home, and (ii) in industry?
b What is the advantage of detergent over soap in a hard water area?
c Why is it more expensive to live in a hard water area than in a soft water area?

Softening water

To soften water you need to get rid of the calcium and magnesium ions. You can do this by adding sodium carbonate. The calcium and magnesium form insoluble calcium carbonate and magnesium carbonate, which precipitate out of solution, leaving soft water.

You can also soften water by using an ion exchange resin. The hard water flows through a column that has been filled with tiny beads of the ion exchange resin. These beads have hydrogen or sodium ions on their surface. The calcium and magnesium ions in the water are attracted to the beads and are replaced in the water by sodium or hydrogen ions. Most water softeners exchange the calcium and magnesium ions for sodium ions. You then have to put the sodium ions back onto the beads by washing through the ion exchange column with a concentrated solution of sodium chloride (salt).

Questions

d Write a balanced symbol equation, with state symbols, to show what happens when water is softened by precipitation.
e What happens to the concentrations of (i) sodium ions, (ii) calcium ions, and (iii) magnesium ions when you soften water using an ion exchange column that is washed through with salt?
f What would happen to the pH of water that was softened by using a hydrogen ion exchange column? Explain your answer.

Key points

- In some areas, water dissolves calcium and magnesium compounds as it seeps through soil and rock. The water is said to be 'hard'. In other places the water is said to be 'soft'.
- If hard water is heated, insoluble calcium and magnesium compounds form and stick to heating elements and other surfaces.
- The ions in hard water react with soap and detergents, leaving scum round baths and sinks.
- You can soften water by adding sodium carbonate or using an ion exchange resin. This removes the calcium and magnesium ions from the water.

Scientific findings

A Research in Taiwan in 1996 showed that deaths from heart disease in soft water areas were 9.6% higher than in hard water areas.

B Researchers in England in 1981 found that there were 10–15% fewer deaths from heart disease in areas with medium-hard water than in areas with soft water. They did not find even fewer deaths in areas with very hard water.

C In 1975 the Public Health Council in the Netherlands came out against softening water as part of the water treatment process. This was because studies between 1958 and 1970 had shown that there were lower death rates from heart disease in people from 23 communities supplied with hard water than in those supplied with soft water.

D Men suffering their first heart attack between the ages of 35 and 74 were the subject of a careful study of 18 946 men in Finland. It showed that a one-unit increment in water hardness decreased the risk of the first heart attack by 1%.

E Deaths from cancer of the rectum (986 cases) of Taiwan residents from 1990 through 1994 were compared with deaths from other causes (986 controls). There was a statistically significant negative relationship between drinking water hardness and rectal cancer mortality.

F A 1988 report describes a study of 7500 primary and secondary school kids in Britain. Episodes of eczema over a one-year period were 17.3% higher in areas with very hard water and 12.0% higher in areas with moderately hard water. This effect was seen only in the primary-age children.

G In 1997 the findings of a study in Taiwan were published. There was a significant negative relationship between drinking hard water and death from gastric cancer.

H In 2004, research was published that linked drinking softened water and gum disease (gingivitis).

I In a study of 18 patients published in 1999, drinking hard water was linked to changes in the urine that indicated that calcium oxalate might be building up in the kidney. This compound causes kidney stones.

J Soft water was hardened by adding calcium ions in a study published in 1993. There was a fall in the lead levels in the water, and an even greater drop in the lead levels of the people drinking it, compared to similar areas where the water was not hardened. Lead is known to cause brain damage.

K Research published in 1986 measured life span and the magnesium and calcium contents of drinking water for specific regions of the USA. The data show a strong positive correlation between low magnesium content and decreased life span, and between high calcium and magnesium content and increased life span.

Making sense of it

Questions

a What is meant by a 'positive correlation' or 'positive relationship'?

b What is meant by a 'negative correlation' or 'negative relationship'?

c (i) Which pieces of research suggest that drinking hard water has a good effect on health?

(ii) List these good effects.

d (i) Which pieces of research suggest that drinking softened water has a negative effect on good health?

(ii) List these negative effects.

A few more facts

Doctors think that the intake of salt (sodium chloride) is linked to heart disease. The increased sodium ion concentration in the blood causes increased blood pressure, which makes it more difficult for the heart to pump the blood around the body.

In a study published in 1997, 59 water samples that had passed through home water softeners were tested for sodium ions. The non-softened water had 110 mg/litre sodium ions. The softened water had at least 278 mg/litre, and 17% of the households had sodium ion levels of over 400 mg/litre. This would mean that a person from one of these 17% of households would consume 33% of the recommended daily limit of sodium just from the tap water.

Questions

e Someone you know is trying to decide whether they should put a water softener in their home. The salesman has pointed out the advantages of having a water softener, like less limescale, better energy efficiency and using less detergent. What would you recommend and why?

f You are living in a house where the water is softened because there is a child with severe eczema. What could you do to decrease your sodium ion intake?

Key points

- Some issues are not clear cut. Whether we should soften water is one of these issues.
- Evidence for or against a course of action can be drawn from many scientific studies across the world.
- We need to be able to consider all the evidence and come to a balanced view.
- The final decision may depend on the particular needs or vulnerabilities of the people affected.

C3 4 Energy and reactions

Chemical reactions in your body release energy. This keeps you warm, lets you see and feel, moves your muscles, allows your brain cells to communicate, and enables the thousands, if not millions, of other processes that keep you alive. The same is true for the life processes of all organisms, from the tallest tree to the smallest bacterium.

The reactions of life

Look at the athlete. Her muscles are contracting, pushing her limbs against the ground to produce kinetic energy. Where does this energy come from? It comes from food, the fuel of the body, and oxygen. Oxygen and digested food, the raw materials for respiration, are transported to every cell. Respiration is a series of chemical reactions, the step-by-step oxidation of food that releases energy. This energy is used to 'charge' the chemical battery of the cell, an amazing substance we call ATP. Wherever energy is needed, the ATP reacts, releasing the required energy.

Fuelling history

Our devices to use chemical reactions to transfer energy have a long way to go before approaching the incredible versatility and elegance found in nature. Humans started using chemical reactions to transfer energy about 1.5 million years ago, when *Homo erectus* learnt the skill of making fire. Cooking food was probably the earliest example of our use of chemical reactions, followed by firing ceramics (25 000 BC), smelting gold and silver (4000 BC), creating bronze (3500 BC) and making iron (1000 BC). All these processes require energy and need a suitable fuel to be burned.

What fuels were used? We assume that early peoples used wood or maybe charcoal for fuel, but the earliest stone lamps, probably burning animal fats, date from 70 000 BC. Jumping forward, the Chinese used natural gas for fuel as early as 900 BC and had started to use coal by 300 AD. Fossil fuels were not used in Europe until much later.

Propulsion

By 850 AD the Chinese had invented a primitive version of gunpowder, and by 1050 recipes for three different forms had been published. A hundred years later the Chinese were using gunpowder to power rockets, and by 1220 they were using bombs in warfare. By 1390 gunpowder was being used for warfare in Europe.

▲ Wan Hu's flight.

Some early scientists did see the potential of explosive, energy-releasing reactions for travel rather than as weapons. In 1500 a Chinese scientist, Wan Hu, tied 47 gunpowder rockets to the back of a chair in an attempt to build a flying machine. Unfortunately the device exploded, killing Wan, who was acting as test pilot. Travel is more reliable with controlled explosions, like the tiny one that happens every cycle in the internal combustion engines that power our cars. However, the huge explosions that propel modern rockets into space are rather reminiscent of Wan Hu's experiment.

The future

Once Britain was covered with forest. Most of the trees ended up in the charcoal burners' pits, used to fulfil people's demands for fuel. Today we have fossil fuels but these will soon be gone, like the forests. The search is on for new fuels, and the hope is that the fuels of tomorrow will be cleaner and more effective than the fuels of today.

Think about what you will find out in this section

How to measure the energy stored in foods and fuels.	Exothermic and endothermic reactions.
Using bond energies to estimate energy changes.	The challenge of finding new fuels.

How much?

How much fuel will I need to heat my house or travel 100 kilometres in my car? How much food do I need so that I do not put on weight or lose it? How much ammonium nitrate should you include in an 'instant cold pack'? How much explosive do you need to put in a single bullet?

You already know that some chemical reactions are exothermic (give energy out) and others are endothermic (take energy in). To answer the questions above, you need to know more. You need to know how much energy is given out or taken in during the chemical reaction.

▲ How much explosive is needed to propel a bullet?

Burn it

Food labels show nutritional information, including the amount of energy in 100 g of the food. This number comes from measurements made using a **bomb calorimeter**. A sample of the food is burned in oxygen inside the 'bomb'. The energy released is transferred to water, causing the temperature to rise. This temperature rise is measured, and the scientist works out how much energy was released, in joules.

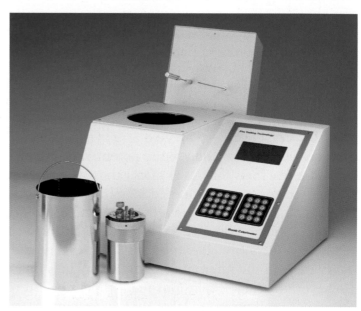

▲ A bomb calorimeter.

Questions

a The walls of the 'bomb' are usually made of stainless steel. Suggest why this is a good choice.

b The water is stirred. Explain what might happen if the water were not stirred.

c There was 500 cm³ of water in the calorimeter with a starting temperature of 20 °C. The final temperature was 67 °C. It takes 4.2 J to heat 1 cm³ of water by 1 °C. What is the minimum energy given out in the reaction?

d Why is the answer to **c** the minimum amount of energy given out?

Even though the calorimeter is really well insulated, not all the energy ends up in the water. Some of the energy heats up the 'bomb' itself, and there will always be some heat loss to the surroundings. To allow for this, a bomb calorimeter is **calibrated**. The scientist works out the temperature rise of the whole calorimeter for a known amount of energy.

The calorimeter shown is calibrated using benzoic acid. Burning 1.000 g of benzoic acid releases 26.429 kJ of energy. If burning 1.000 g of benzoic

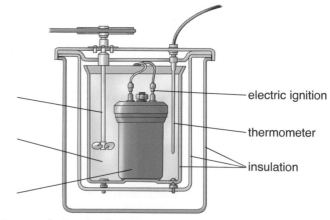

▲ Diagram of a bomb calorimeter.

electric ignition

thermometer

insulation

acid makes the temperature of the calorimeter go up by 23.76 °C, then you know that it takes 1.112 kJ to raise the temperature of the calorimeter by 1 °C.

e *The calorimeter described above was used to test a food for its energy content. Burning 1.000 g of the food caused a temperature rise of 23.91 °C. What is the energy content of the food per 100 g?*

▲ An 'instant' cold pack.

Cup calorimeter

'Instant' cold packs contain separate containers of solid ammonium nitrate and water. When needed, the pack is flexed to break the containers inside and the two substances mix. An endothermic reaction happens and the temperature of the pack drops, so that it can be used instead of an ice pack. The temperature fall should not be too great, or a cold burn may be added to the patient's injuries. A drop in temperature of about 20 °C is ideal. Some students were asked to find out how much ammonium nitrate should be put into an 'instant' cold pack.

Two students decided to use a simple **calorimeter** made from a polystyrene cup, like the one shown in the photo. They looked up the solubility of ammonium nitrate and discovered that it was 214.4 g per 100 g water. They put 200 g of water in the polystyrene cup, took the temperature and added 35 g of finely powdered ammonium nitrate. They then swirled the cup gently and measured the minimum temperature. They discovered that the temperature of the mixture had dropped by 10.4 °C.

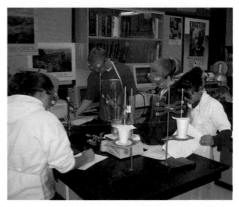

▲ A simple calorimeter.

Questions

f *Why did they use a polystyrene cup?*
g *Why did they check the solubility of the ammonium nitrate?*
h *Why did they use finely powdered ammonium nitrate?*
i *The students want a temperature drop of about 20 °C. Suggest two different ways of achieving this, decide which you think is better, and explain why.*

Energy conventions

Changes in energy are given the symbol △**H**. The △ means change and the H stands for **enthalpy**, a scientific term for energy. Energy given out in an exothermic reaction, like burning, is given a negative sign. Energy taken in in an endothermic reaction, as in the instant cold pack, is given a positive sign.

Key points

- A calorimeter is used to find out the amount of energy given out or taken in during a physical or chemical change.
- A calorimeter can be used to measure the energy given out when a substance is burned. The released energy is transferred to water, and the temperature rise is used to calculate the energy released.
- It is important to minimise energy losses to the environment when carrying out calorimetry experiments.
- Changes in energy are shown by the symbol △H. Endothermic reactions are represented by +△H. Exothermic reactions are represented by −△H.

First principles

You have to remember that energy is conserved. Energy cannot be created or destroyed: it is just moved about. When you eat, the energy in the food is going to end up somewhere. Perhaps it will show up as movements of your body, as thermal energy, as a signal travelling down a nerve, or stored in a fat cell.

Tracking that energy

Many foods have labels showing nutritional information. The picture shows a sample label given by the food standards agency. It tells us a lot about the food, including the energy. As you can see, each 100 g contains 58 kcal (kilocalories). The original 'calorie' is an older unit of energy than the joule, and equals 4.2 joules (J). The energy content is also shown as 245 kJ (kilojoules).

The 'right' amount of energy for a person to take in depends on what he or she does. A professional cyclist will need a lot more energy than a committed couch potato. The **recommended dietary allowance (RDA)** of energy for men and women is shown in the table on the right. Of course, this does not tell you exactly what energy you need (that varies from individual to individual), but it is a guide.

▲ Nutritional information.

Question

a Suggest how the energy content of the food was measured.

Questions

b What is the RDA for energy in kilojoules for a 15–18-year-old (i) female, and (ii) male?
c Why does the table suggest +500 kcal for a lactating female?

Many of us take in too much energy for the amount of exercise we do. We run the risk of becoming overweight or even obese. In the table below, look at the nutritional information for a banana and a packet of crisps. Mass for mass, banana has only 20% of the energy in crisps.

Food	Mass of portion (g)	kcal in 100 g	Mass in 100 g of the food			
			Carbohydrate	Protein	Fat	Fibre
Banana	150	95.0	23.2	1.2	0.3	1.1
'Lite' crisps	28	470.0	60.0	7.5	22.0	5.0
Butter	10	737.0	0.0	0.5	81.7	0.0
Sugar	6	417.0	99.8	0.0	0.0	0.0
Cod	150	105.0	0.0	23.0	0.9	0.0
Celery	50	14.0	3.5	0.7	0.2	1.6

Age	Energy (kcal)
Males	
11–14	2500
15–18	3000
19–24	2900
25–50	2900
51+	2300
Females	
11–14	2200
15–18	2200
19–24	2200
25–50	2200
51+	1900
Pregnant	+300
Lactating	+500

▲ Recommended dietary allowance (RDA).

Now look at the nutritional information for butter, sugar and a white fish like cod. Butter is almost all fat, and 100 g of butter gives 737 kcal. 100 g of sugar gives 417 kcal. Cod is almost all protein and gives 105 kcal per 100 g. Fats and oils are the most energy-rich food group, followed by carbohydrates, so controlling our fat intake is good way of controlling our weight. The British Heart Foundation suggests that a food that contains 20 g of fat per 100 g of food is a high-fat food, while one with 3 g of fat is a low-fat food. This is why you see so many adverts saying 'less than 3% fat'.

Does the route matter?

Respiration and combustion have the same reactants, sucrose and oxygen, and the same products, carbon dioxide and water. The same amount of energy should be released by both respiration and combustion.

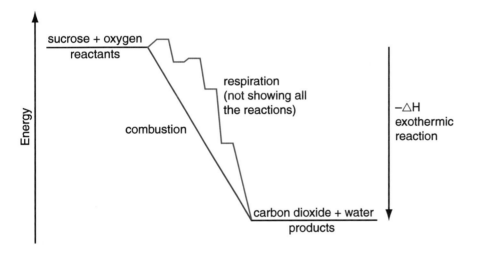

▲ Energy level diagram.

The energy level diagram compares where the energy goes in the two processes. Combustion occurs in one reaction (the red line), while respiration happens in many reactions (the blue line). Overall the same amount of energy will be released. This is an application of the law of conservation of energy. It confirms that the energy content of sugar measured using a bomb calorimeter, 16.48 kJ/g, is also the energy released from 1 g of sugar during respiration.

Even so, the *useful* amount of energy our bodies get from 1 g of sugar may vary. Our cellular processes convert only about 40% of the energy in sugar into a chemical store that our cells can use. The rest is released as thermal energy. We also expend energy eating and digesting food, then transporting it to our cells. Scientists have calculated that we expend more energy eating celery than we get from the celery we are eating. This is because celery is mainly fibre, water and some tasty molecules – the perfect diet food!

Question

d A friend argues, correctly, that there is more energy in a banana than in a packet of 'lite' crisps. Do you agree with her that eating the packet of crisps is healthier than eating the banana?

Question

e What mass of celery contains the same amount of energy as a packet of 'lite' crisps?

Key points

- Energy is conserved. When you eat, you take in energy from your food to keep your body going. If your body does not need all the energy, it will store the extra as fat.
- Nutritional information tells us how much energy our food contains. The amount is shown in calories and kilojoules.
- Proteins contain less energy per gram than fats, oils and carbohydrates.
- Some foods are mainly fibre and water – they are low in calories.

Breaking bonds

Think about a crystal of salt. It contains millions of sodium ions, Na⁺, and an equal number of chloride ions, Cl⁻. The positive and negative ions attach to each other strongly. We say there are strong **ionic bonds**. We would have to put in a lot of energy to drag the positive and negative ions apart. Breaking bonds needs an input of energy.

Making bonds

What happens if we reverse the process shown in the diagram and allow separated sodium and chloride ions to come together? They would bond together and form solid sodium chloride. In the process energy would be released. The same amount of energy would be released as was taken in when the bonds were broken.

Bonds and energy

This applies to all types of bonds. You need to put in energy to break bonds: it is an *endothermic* process. You get energy out when you make bonds: it is an *exothermic* process. Most people find it easier to remember that breaking bonds needs energy. You can think of the atoms being ripped apart! Then think of making bonds being the reverse.

It is useful to know the amount of energy associated with each type of **covalent bond**. Scientists think that about 413 kJ of energy is associated with one mole of C—H bonds, so if you make one mole of C—H bonds you release about 413 kJ, and to break one mole of C—H bonds you have to put in about 413 kJ. They have worked out these **bond energies** by doing many calorimetry experiments. Using these bond energies, you can estimate how much energy will be given out or taken in during a reaction involving covalent compounds.

Burning methane

Let's start with a familiar reaction. How much energy do we expect to get out when we burn one mole of methane in an excess of oxygen?

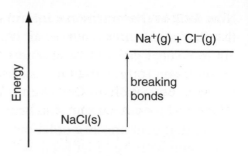

▲ Energy level diagram for breaking bonds in sodium chloride.

Question

a Why is the same amount of energy given out when the ions come together as was taken in when the ions were pulled apart?

Question

b Explain why it may be useful to be able to estimate the amount of energy being released during a reaction before allowing the reaction to take place.

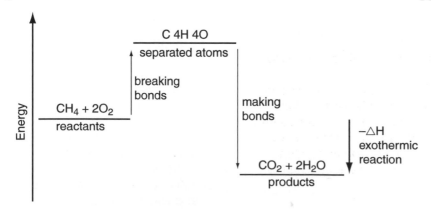

▲ Energy changes involved in burning methane.

```
                H
                |
methane   H — C — H
                |
                H

oxygen    O = O

carbon dioxide   O = C = O

water   H — O — H

                H   H
                |   |
ethane   H — C — C — H
                |   |
                H   H
```

▲ Structures of molecules.

The diagram shows the method of calculation we are using. First we break all the bonds. Look at the structures of the molecules. This allows us to work out what bonds are broken. There is one molecule of methane, containing four C−H bonds. There are two molecules of oxygen, each with an O=O bond. The table tells us the bond energies. The energy we need to put in is:

$$(4 \times +413) + (2 \times +498) = +2648 \text{ kJ per mole of methane}$$

This is shown as *positive*, because we have to put energy in. Then we make the new bonds. We make one molecule of carbon dioxide, which contains two C=O bonds. We also make two molecules of water, each of which contains two H−O bonds. The energy we get out is:

$$(2 \times -805) + (4 \times -464) = -3466 \text{ kJ per mole of methane}$$

This is *negative*, because we are getting energy out. Then we put the two together, giving the overall energy change:

$$\triangle H = +2648 + -3466 = -818 \text{ kJ per mol}$$

When we look up the 'real' answer, from burning methane in a bomb calorimeter, it is −890.3 kJ/mol. The estimate using bond energies is just that, an estimate, but it gives a good guide.

In this example, more energy was given out when the new bonds were made than was taken in when the old bonds were broken, so it was an exothermic change. If less energy was given out when the new bonds were formed than was taken in to break the old bonds, it would be an endothermic change.

Bond	Average bond energy (kJ/mol)
C−H	413
O=O	498
C=O	805
H−O	464
C−C	347

Questions

c Use bond energies to estimate the amount of energy released when you burn one mole of propane (C_3H_8) in an excess of oxygen.

d Look at the table showing the bond energy of C=O in different molecules. Suggest why working out the energy released or absorbed in a reaction using bond energies can only be an estimate.

Bond	Molecule	Bond energy (kJ/mol)
C=O	O=C=O	805
C=O	H−C=O with H below C	695
C=O	H−C−C−C=O (with H's on first two carbons)	736
C=O	H−C−C−C−H (with O double bonded to middle C, H's around)	749

Key points

● Energy is needed to break bonds. Energy is given out when new bonds are made.
● In exothermic reactions, more energy is given out when the new bonds were made than was taken in when the original bonds were broken.
● In endothermic reactions, less energy is given out when the new bonds were made than was taken in when the original bonds were broken.
● Scientists have worked out a 'bond energy' for each type of covalent bond. We can use these bond energies to estimate the energy change of a reaction.

The challenge

What makes a good fuel? It depends on the purpose for which the fuel will be used. It is a particularly difficult challenge to find suitable fuels for rockets, the vehicles that carry satellites, telescopes, space probes and occasionally humans into space.

Like all fuels, rocket fuels need to release energy when they react. The products of this reaction must be gases because rockets work by throwing waste gases out at very high velocities, pushing the rocket in the opposite direction. Like other fuels for vehicles, it is convenient if the rocket fuel is a liquid and can be pumped into the fuel tanks. Rocket fuel has to be lightweight, to minimise the mass of the rocket being lifted. Unlike the fuels we use on Earth, rocket fuel cannot react with oxygen from the air: if oxygen is needed it must be carried by the rocket. **Monopropellant** fuels are particularly suitable for rockets. These are fuels made up of a single substance that releases energy when it decomposes. If the fuel is a monopropellant, only one fuel tank is needed and mixing problems do not arise.

▲ The Space Shuttle steers using rockets.

Question

a List the properties of the ideal rocket fuel.

Hydrazine

Hydrazine, N_2H_4, has been developed as a rocket fuel. Hydrazine has a boiling point of 114 °C and a melting point of 2 °C. It has a relative molecular mass of 32 and a density of 1.02 g/cm^3. Hydrazine decomposes as shown in the equation below. The **state symbols** in the equation tell us the state of matter of the reactants and products, which is particularly relevant when discussing rocket fuels.

$$3N_2H_4(l) \rightarrow 4NH_3(g) + N_2(g)$$

Is the decomposition of hydrazine exothermic? We can use bond energies to estimate the energy change. Look at the display formulae of the molecules involved. Each hydrazine molecule contains one N–N bond and four N–H bonds. The energy needed to break these bonds is $1 \times +158 + 4 \times +391 = +1722$ kJ. There are three hydrazine molecules in the equation, so +5166 kJ is needed.

Each ammonia molecule contains three N–H bonds, and four molecules are made for every three molecules of hydrazine reacting. The energy released from making the ammonia is $3 \times 4 \times -391 = -4692$ kJ. One molecule of nitrogen gas is also made, releasing another –954 kJ. This makes a total of –5646 kJ. We can now work out the energy change:

$$\triangle H = +5166 + -5646 = -480 \text{ kJ}$$

Remember that this energy change is for three moles of the fuel – the equation shows three moles of hydrazine reacting. We always express changes in energy 'per mole', so we must divide the energy change

hydrazine H — N — N — H
| |
H   H

ammonia H — N — H
|
H

nitrogen N ≡ N

hydrogen H — H

Bond	Average bond energy (kJ/mol)
N–H	391
N–N	158
N≡N	954
H–H	436

Question

b One of the problems with hydrazine is that the ammonia produced can react again:

$2NH_3(g) \rightarrow N_2(g) + 3H_2(g)$

Estimate the energy change for this reaction and suggest why this second reaction is a problem.

by 3. The estimated energy change for the reaction is −160 kJ/mol, so the decomposition reaction is definitely exothermic.

Is it reliable?

If the decomposition of hydrazine is exothermic, why doesn't it spontaneously explode? The answer is that the molecules need to have a certain amount of energy before they can react – the **activation energy**. For this reaction, the activation energy is high enough that the hydrazine can be manufactured, transported, pumped and stored without exploding.

So how do you get it to react? The answer is by using a catalyst. Catalysts speed up reactions by lowering the activation energy of the reaction. In this case, the catalyst of choice is the transition element iridium. To make the reaction happen, the hydrazine is pumped through heated ceramic beads coated with iridium. The hot gases made in the reaction explode out as the exhaust of the rocket.

Use it?

Hydrazine is hazardous stuff, as you can see from the table at the right. It is also expensive. When it was first suggested as a rocket fuel in 1959 NASA thought it would cost less than £1/kg to produce in large quantities. In fact, because of the necessary safety precautions and the increased regulation of environmental damage, it costs about £10/kg to make. Compare this to the cost of petrol at just over £1/kg. Even so, hydrazine is used as a monopropellant in situations where its advantages outweigh these disadvantages. It is used in the steering rockets of the Space Shuttle and to power the emergency power unit of the F16 fighter plane.

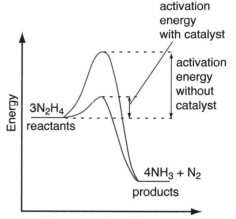

▲ Progress of reaction with and without catalyst.

Question

c (i) Why is the iridium made into beads?
(ii) Suggest why the beads are made of metal-coated ceramic and not solid iridium.
(iii) Look carefully at the diagram above. Suggest why the beads need to be heated.

Safety information for hydrazine

Ingestion	Extremely toxic, possibly carcinogenic
Inhalation	Very dangerous – extremely destructive to upper respiratory tract
Skin	Can cause severe burns, be absorbed into bloodstream
Eyes	Can cause permanent damage
LD_{50}	As low as 25 mg/kg
LD_{50} is the dose that will kill 50% of experimental animals	

Question

d Use the information on these pages, including the safety information in the table, to write an evaluation of hydrazine as (i) a rocket fuel, and (ii) a fuel for ground vehicles.

▲ F16 fighter plane.

Key points

- Exothermic reactions are often not spontaneous. This is because they have a high activation energy.
- Activation energy can be reduced using a catalyst. This may make the reaction spontaneous.
- When evaluating a fuel, many factors need to be considered. Fuels that are suitable in one context may be unacceptable in another.

Alternative fuels

The most commonly used fuels today are coal, natural gas and fuels made from oil. These are non-renewable fossil fuels, and reserves of them are running out. Worse, we now believe that adding carbon dioxide to the atmosphere is causing global warming: all these fuels produce carbon dioxide when they burn. What are the alternatives? One is to use biofuels, which release the same amount of carbon dioxide when they burn as they did when they grew. The other is to use fuels that contain no carbon, like hydrogen.

Enough energy?

Would these alternative fuels release enough energy? We need to compare the energy given out by different fuels using their **energy values**. The energy value is the amount of energy released when 1 g of the fuel is burned. It is measured using a bomb calorimeter (see also section 4.1). The table shows the energy value of 11 fuels. The runaway winner appears to be hydrogen, with a huge energy value of 142 kJ/g.

Fuel	Fossil fuel	Energy value (kJ/g)
Biodiesel		41
Bioethanol		30
Biomethane		56
Coal	✓	10
Diesel	✓	42
Fuel oil	✓	44
Hydrazine		5
Hydrogen		142
LPG	✓	50
Natural gas	✓	60
Petrol	✓	48

Question

a How does the energy value of biofuels compare to fossil fuels?

The wonder fuel?

When you burn hydrogen you make only water; no carbon dioxide, no sulfur dioxide, no carbon particles, no unburnt hydrocarbons. It seems the ideal, pollution-free fuel, so why isn't it used more widely?

First, it is a gas with a very low density. 1 g of hydrogen has a volume of 12 litres. Think about a car with a fuel tank of 60 litres. Filled with hydrogen, that would contain 710 kJ = 0.71 MJ. Petrol is a liquid and much denser. The same fuel tank filled with petrol would contain 19 353 MJ. Suddenly hydrogen does not look so attractive! Also, handling a gas is much more difficult than handling a liquid. Gases are harder to store and more difficult to pump from one place to another. Second, hydrogen has a reputation for being dangerous. In 1937 the most famous German passenger airship, the Hindenburg, made its first trans-Atlantic flight. As it came into land, it burst into flames and crashed, giving hydrogen its bad repuation.

▲ The Hindenburg airship.

In fact, hydrogen is no more dangerous than petrol, and the other problems can be addressed by supercooling and pressurising the hydrogen to make it liquid. The 60 litre tank would then contain 596 MJ worth of liquid hydrogen fuel, compared to 19 353 MJ for the same tank of petrol.

Some car manufacturers have already solved the problems of using hydrogen as a fuel. The Mini shown in the photo has a conventional internal combustion engine that runs on liquid hydrogen.

Another approach is to use hydrogen as a fuel for a **fuel cell**. Cars running on fuel cells would have electric motors, like cars running on batteries. Batteries produce electrical energy from chemical reactions, but the reactants run out. The battery becomes 'flat' and has to be discarded or recharged. In a fuel cell you have a constant supply of fresh reactants, for example oxygen from the atmosphere and hydrogen from a fuel tank. A number of car manufacturers have already made cars that run on fuel cells, but these cells are still far too expensive, because they are in the development phase and each has to be made as a 'one-off'. As the technology develops and more are made, costs will come down.

▲ This Mini has been converted to run on liquid hydrogen.

▼ This hydrogen fuel cell prototype car, the GM HydroGen3, cost $1 million.

Questions

b Write a balanced equation for the combustion of hydrogen.

c Liquid hydrogen and hydrogen gas have the same energy value, but liquid hydrogen is a much more practical fuel. Explain why.

d Suggest why cars that run on fuel cells may be further into the future than cars powered by hydrogen combustion.

Key points

- Reserves of fossil fuels such as coal, natural gas and oil are running out. We need to look for alternatives.
- The energy value of a fuel is the energy released when 1 gram is burned.
- We can compare the energy released by different fuels using their energy values.
- Hydrogen has a high energy value and does not produce unwanted waste gases when it burns.
- Hydrogen can be burned or used as a fuel for fuel cells.

The problem

The purpose of biofuel projects is to make fuels from crops. In Brazil, ethanol is made from sugarcane and used in place of petrol. In Germany, oilseed rape is used to made biodiesel. There is even a power plant in Suffolk that uses chicken droppings as an energy source! Looking at projects across the world helps us decide whether biofuels are a practical alternative to fossil fuels.

Conventional crops

When we think about biofuels, we think first about the energy value of the fuel, as discussed in section 4.5. We also have to consider what resources were needed to produce the biofuel. How much fertiliser? How much fuel in the tractors and harvesters? How much energy to turn the biomass into a liquid fuel? The table shows a thorough study of the problem. The scientists have looked at the energy, other than solar energy from the Sun, put in to grow, harvest and process the crops to make biofuels. They looked at two different methods. In the first, the straw left when the oilseed rape or wheat was harvested was ploughed back into the ground. In the second, the straw was also harvested, and then used as a fuel. All the amounts are given in megajoules per hectare (MJ/ha). One hectare is $10\,000\,\text{m}^2$.

▲ Poultry litter is used to generate electricity.

	Energy yield (+) or cost (−) (MJ/ha)			
	Oilseed rape, straw ploughed	Oilseed rape, straw utilised	Wheat, straw ploughed	Wheat, straw utilised
Biodiesel / bioethanol	+54346	+54346	+74189	+74189
Cake / bran	+1316	+1316	+0	+0
Straw	+0	+60000	+0	+97500
Total	+55662	+115662	+74189	+171689
Agricultural fuel	−4687	−4945	−4300	−4773
Fertilisers	−7190	−7190	−7815	−8070
Agrochemicals	−337	−337	−1045	−1045
Seed	−35	−35	−925	−925
Packaging	−282	−282	−447	−485
Transport	−723	−1122	−1495	−2149
Processing	−17251	−17251	−50810	−50810
Total	−30505	−31162	−66837	−68257
Net energy gained	+25157	+84500	+7352	+103432

▲ Energy balance for biodiesel from oilseed rape and bioethanol from wheat.

Questions

a How much energy was stored in biodiesel for every hectare of oilseed rape grown?

b How much energy was stored in bioethanol for every hectare of wheat grown?

c How much energy was released by burning the straw from one hectare of oilseed rape?

d How much energy was released by burning the straw from one hectare of wheat?

e Calculate the net energy gain as a percentage of the energy stored in the biofuel for
(i) biodiesel with straw ploughed in
(ii) biodiesel with straw used as fuel
(iii) bioethanol with straw ploughed in
(iv) bioethanol with straw used as fuel.

f What do these data tell you about
(i) producing liquid biofuels that can be used in cars?
(ii) using waste biomass like straw as a fuel?

Waste not, want not

Biodiesel and bioethanol production is much more efficient if the straw, the waste product, is used for fuel. The power plant in Suffolk uses poultry waste to produced methane. We throw away a lot of material that could be rotted to make methane, or just burned to release thermal energy.

New crops

One group of American scientists has suggested growing algae in huge, shallow, saltwater pools in desert areas in California. The pools would be made by diverting the water that drains from nearby agricultural areas that have irrigation systems. This water contains fertilisers that have been polluting lakes and seas, disrupting the balance of their ecosystems. The algae would be harvested from the pools and used to make biodiesel. The scientists estimate that the total fuel requirement of the USA could be provided using 4 million hectares of desert. This sounds a lot, but to put this in perspective a single desert in California, the Sonora, covers 31 million hectares.

▲ At the moment we see algae as a problem, but perhaps they could be a solution.

Hydrogen from biomass

Hydrogen is a fuel with many advantages (section 4.5). Scientists have come up with a way of making hydrogen from biomass. If biomass is heated under very controlled conditions with only a limited amount of oxygen, incomplete combustion occurs. Carbon monoxide and hydrogen are produced. The mixture can then be used as a fuel, or separated into hydrogen and carbon monoxide for separate use.

Question

g Make a list of all the waste materials that could be
(i) burned to release thermal energy, and
(ii) fermented to make biomethane.

Questions

h What raw materials and energy source are needed for this project to work?

i Why is a desert area ideal for growing algae?

j Other than replacing fossil fuels with renewable fuels, what environmental problem does this project try to solve?

k Do you think this idea would be suitable for other places? Give your reasons.

Key points

- Scientists are coming up with many different alternatives to fossil fuels.
- Each suggestion needs to be carefully analysed to see if it is practicable.
- Many of these suggestions are based on biomass. Biomass can be burned, processed to make biofuels or rotted to make methane.
- We need to consider waste biomass as well as that grown for the purpose of making fuels.

Questions

l What do you see as the advantages and disadvantages of this fuel?

m Do you think biomass and biofuels will fulfil all our future energy needs? List all the arguments for 'yes' and all the arguments for 'no' before coming to your decision and writing your reasons.

Analyse and identify

In chemistry laboratories we find substances in bottles with labels on them. It is very different out in the real world, where almost all materials are complex mixtures of unidentified substances. Scientists face the challenge of identifying these substances and measuring how much of each is present in the mixture. This is true of scientists in many fields, including medicine, forensic science, environmental studies, geology and cosmology.

Using your senses

Scientists have been identifying substances using their senses for thousands of years. Alkalis feel soapy. Acids taste sour. Ammonia has a distinctive smell. Metals are shiny. Gold is yellow. Many of the modern ways of identifying substances depend on colour. Universal indicator changes colour, helping us to decide whether a substance is acidic or alkaline. Chlorine is pale green and bleaches damp litmus. Compounds containing Cu^{2+} ions are green or blue. Iron compounds are pale green or brown.

But senses alone are not enough. Many substances are colourless. Smelling, tasting or touching unknown substances is not safe. Scientists go beyond the limits of their senses by building machines to detect things they cannot see, hear, taste, smell or touch.

Light from the stars

In 1858, Pierre-Jules-César Janssen, a French astronomer, was examining the Sun during a solar eclipse. He observed the light from the corona, the outer part of the Sun not covered by the Moon. He split up the light from the corona using prisms. He saw a **solar spectrum** like the one in the picture below. It is a kind of graph showing how much light there is at each wavelength.

You can see that there are fine black lines in this spectrum. These wavelengths of light were missing because they were being **absorbed** by elements. The atoms were taking up a noticeable proportion of energy at that particular wavelength. Scientists now know that this is because electrons take in the energy and jump to a higher energy level in the atoms. For example, hydrogen absorbs red light of wavelength 656 nm (1 nm = 1 nanometre = 0.000001 mm) and there is a black line, a missing wavelength of light, at exactly this point in the solar spectrum.

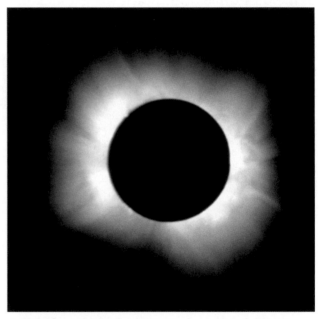

▲ Light from the Sun's corona can be analysed during a solar eclipse.

▲ The solar spectrum of our Sun.

Janssen found a new black line at 587.49 nm, in the yellow part of the spectrum. Another astronomer, Norman Lockyer, realised that this line did not correspond to the light absorbed by any known element and suggested that a new element had been found. He named the unknown element helium, from *Helios*, the Greek name for the Sun. In 1882, helium was found on Earth, erupting from the volcano Vesuvius. The study of spectra opened the door for the discovery of all the Group 0 elements.

Once an element has absorbed energy, that energy can be given out again, or **emitted**. We now know this is because the electrons fall back to their original energy levels, giving out the energy they absorbed. If helium absorbs light of wavelength 587.49 nm, it should also give out light at the same wavelength. The colours of light given out when helium is heated are shown in the photo below. As you can see, there is a very distinct yellow line.

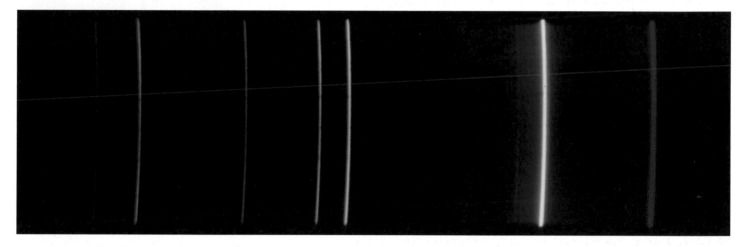

▲ A spectrum of the light given out by helium.

Colours of light

Light and the other parts of the electromagnetic spectrum, like ultraviolet and infrared radiation, are very important to chemists. When matter interacts with electromagnetic radiation it takes in energy. When energy is given out by matter it often leaves as electromagnetic radiation. Sometimes we analyse this interaction with our eyes. Sometimes we use scientific instruments, like Janssen and Lockyer who used prisms and telescopes. The study of how matter interacts with electromagnetic radiation is called **spectroscopy**.

Think about what you will find out in this section	
How to test for common ions.	Working out the empirical formulae of compounds.
How to use spectroscopy to analyse matter.	The advantages of modern instruments.
A modern separation technique.	

Quick and easy

When faced with an unknown material, the first question a chemist asks is 'What is it?'. In scientific terms, he or she wants to do a **qualitative analysis**. This tells you what substances are present but not how much of each substance is present, just that it is there.

Modem laboratories contain expensive, elaborate machines that can carry out qualitative analysis. Even so, most chemists have a repertoire of quick and easy tests that they can complete with the minimum of equipment. This can be particularly useful when surveying the rocks of the mountains in Chile, or checking soil quality in Mali, for example.

A starting point

Chemists usually know where the unknown material has come from. It could be **organic**, a carbon-based compound, or **inorganic**, which refers to everything else. If unsure, the chemist would carry out a very basic test. If you heat organic compounds in air they turn black, or char, because the organic matter begins to burn, producing carbon, just as when wood is burned in a fire.

▲ Kit for quick and easy chemical tests.

We are going to begin with samples of inorganic materials. Most inorganic solids found in the environment are mixtures of ionic compounds, including a huge range of salts, rocks and minerals. There are, of course, exceptions. Gold is found as a metal. Sand is made from silicon dioxide, a giant covalent compound. The comforting thing about exceptions is that there are only a few of them and the chemist is likely to recognise them if they turn up.

When chemists analyse inorganic compounds, a large part of what they do is to identify which ions are present. There are tests for every type of positive ion and for every type of negative ion. On these two pages we are going to concentrate on tests for a few of the positive ions.

Flame tests for metal ions

The quickest, easiest and most popular test is a **flame test**. Some metal ions give out coloured light when heated. You take a length of wire of an unreactive metal, ideally platinum. You make sure the wire is absolutely clean by dipping it in concentrated hydrochloric acid and burning off any impurities in a bunsen burner flame. You dip the wire into clean concentrated hydrochloric acid. Then it goes straight into the sample. Some of the sample material sticks to the wire. Put the sample into the bunsen flame and observe the colour of the flame.

This method is particularly useful for identifying some Group 1 and Group 2 metal ions. The photos show the distinctive colours associated with lithium, sodium, potassium, calcium and barium. Magnesium ions give out electromagnetic radiation when heated, but not at a wavelength we can see, so there is no colour associated with heating magnesium ions.

Questions

a Why must the wire be made of an unreactive metal?
b Explain why cleaning the wire is so important.

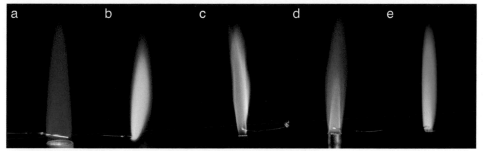

▲ Flame test colours: (a) lithium, (b) sodium, (c) potassium, (d) calcium, (e) barium.

Transition metal compounds – also colour-coded

You already know that many transition elements form coloured compounds. Colours can also be used to identify transition metal ions in compounds. Mix sodium hydroxide solution with a solution containing transition metal ions, and you get an insoluble precipitate. Three of the easiest to identify are copper II hydroxide, which is blue, iron II hydroxide, which is pale green, and iron III hydroxide, which is brown.

Two-stage tests

Some ions cannot be identified in one step. Aluminium hydroxide is white. Like many other hydroxides, it is insoluble in water. It differs from most hydroxides in one important way. If you add more sodium hydroxide to it, the aluminium hydroxide precipitate vanishes. It reacts to form a new, soluble compound. This gives you a way of telling a solution of aluminium ions from a solution of, say, calcium or magnesium ions. When you add sodium hydroxide solution, all three give white precipitates. If you *continue* adding sodium hydroxide solution, only the aluminium hydroxide precipitate vanishes.

Tailor-made tests for particular ions

For some positive ions, you just have to know the test. For example, ammonium ions form ammonia when you add sodium hydroxide. You would probably recognise the smell. You can confirm your guess by putting a piece of damp litmus paper above the surface of the liquid. The ammonia gas will turn the litmus paper blue.

▲ Coloured precipitates: (a) copper II hydroxide, (b) iron II hydroxide, (c) iron III hydroxide.

Key points

- Qualitative analysis reveals what substances are present, but does not tell you how much of each there is.
- Organic substances turn black if you heat them in air.
- Lithium, sodium, potassium, calcium and barium ions can be identified using flame tests. Each has a distinctive colour.
- Adding sodium hydroxide and observing the reaction helps you to identify copper II, iron II, iron III, calcium, magnesium, aluminium and ammonium ions.

Questions

c A solution containing an unknown positive ion gives a white precipitate when sodium hydroxide is added and a brick-red flame test. Which ion is present?

d A solution gives a negative flame test but a white precipitate when tested with sodium hydroxide. The white precipitate redissolves when more sodium hydroxide is added. Which ion is present?

e A solution gives off a pungent gas when you add sodium hydroxide. This gas turns damp red litmus paper blue.
   (i) What is the the gas?
   (ii) What ion was present?

Out and about

Many scientists work in modern laboratories with state-of-the-art equipment to analyse their samples. These machines are very precise and the results they give are absolutely reproducible and very accurate. This is very important when working in a hospital, where a patient's life may depend on the results being accurate, or in a forensic laboratory, where the evidence must be reliable enough to convince a court, or in a water purification plant, where water quality is checked to protect public health.

Not all scientists work in laboratories. Many work 'out in the field'. They can be up mountains, down mines or on a ship looking at samples brought up by divers. They cannot wait until they take their samples back to the laboratory. They may need to make a decision that very day about what to do next. They do the tests they can, using the equipment they have.

As discussed on the previous page, most non-organic solids are made up of ions. Just as there are tests for positive ions, there are tests for negative ions. You will learn about some of the tests for the more common negative ions here.

▲ A scientist working in the field.

Acid makes carbonates fizz

One of the easiest tests is for carbonate ions. You probably know this test already. You drip dilute acid on the sample. If it fizzes, then the sample probably contains a carbonate. The gas being made is carbon dioxide. Confirm your conclusion by carrying out the standard test for carbon dioxide, which is to pass the gas through limewater. If carbon dioxide is present, the limewater will turn milky.

Many rocks contain minerals that are carbonates. Once you know a rock is a carbonate you can use your knowledge of carbonates to identify the mineral. Many carbonates decompose when heated. If you heat a small sample you may see colour changes as the oxide is made. For example, green copper carbonate (malachite) turns into black copper oxide, so a green to black colour change tells you that the mineral was malachite. If the carbonate mineral goes bright yellow when heated, and then white as it cools, it was zinc carbonate: zinc oxide is yellow when hot and white when cool.

▲ (a) Copper II carbonate (malachite); (b) copper II oxide.

Silver nitrate shows up halides

You can test for the presence of the common halide ions with an acidic solution of silver nitrate. Chloride, bromide and iodide will form precipitates of silver chloride, silver bromide and silver iodide. Telling these precipitates apart is slightly trickier, however. Silver iodide is yellow and silver chloride is white. The problem is the silver bromide, which is cream-coloured. On its own, you can be left wondering whether the silver bromide is yellow, or cream, or white. The solution is to make a small amount of silver chloride, silver bromide and silver iodide and compare your 'unknown'. Once you have all three in front of you, the colour difference is easier to see.

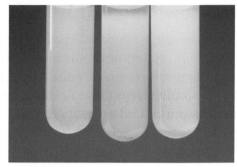

▲ The silver nitrate test for halides, showing (left to right) the colours of silver chloride, bromide and iodide.

Barium reveals sulfate ions

Many sulfates are soluble, but barium sulfate is not. If you mix a solution of sulfate ions with a solution of acidified barium chloride, you will get a white precipitate of barium sulfate.

Aluminium shakedown for nitrate ions

All nitrates are soluble, so you cannot use a 'precipitate test' like the ones for halides and sulfates to confirm the presence of nitrate ions. You have to reduce the nitrate ions to ammonia, and then test for that. You do this by adding some aluminium powder to the solution. The aluminium reduces the nitrate ions to ammonium ions. You then add an excess of sodium hydroxide, and this changes the ammonium ions into ammonia. You can then test for ammonia in the usual way (see section 5.1).

▲ Copper sulfate solution with barium chloride.

Questions

a Limewater is calcium hydroxide solution and the milky precipitate is calcium carbonate. Write balanced symbol equations for the two reactions in the test for a carbonate.
b Write balanced symbol equations for: (i) the decomposition of copper carbonate, (ii) testing for bromide ions, and (iii) testing for sulfate ions.
c Explain how you would test a solution for nitrate ions, including all the experimental details.

Tests in handy packages

If enough people need a test, this makes a market and manufacturers will create a kit. For example, fish owners need to test the water in their fish tanks for a variety of ions, including nitrates. So manufacturers produce kits for the fish owners to use. These test kits can then also be used by scientists in the field.

Key points

- Carbonates react with dilute acids to form carbon dioxide. This gas turns limewater milky.
- An acidic solution of silver nitrate forms a precipitate on contact with chloride ions (white), bromide ions (cream) and iodide ions (yellow).
- If sulfate ions are present, adding acidified barium chloride produces a white precipitate.
- Aluminium powder reduces nitrate ions to ammonium ions. Adding sodium hydroxide to these produces ammonia gas.

Unknown organic

A scientist may need to identify an organic compound as part of a medical, forensic or environmental investigation, or when working out the products of an unexpected reaction during an industrial process. Identifying one of these compounds can be tricky, as organic compounds are alike in many ways.

Revealing reactions

Most organic compounds react in characteristic ways, depending on the groups of atoms they contain. For example, some organic compounds are **unsaturated**. This means that they contain at least one carbon–carbon double bond, C=C. All compounds with this group react with bromine. When you add the organic compound to bromine water, the orange colour of the bromine disappears.

> Question
>
> **b** What would you see when you add (i) propane and (ii) propene to bromine water?

> Question
>
> **a** What test is used to show that an unknown material is an organic compound? (Hint: see section 5.1)

▲ Unsaturated organic compounds decolorise bromine water.

Elemental ratios

So far we know that the unknown compound is organic, and maybe a little about the reactive groups of atoms it contains. The next step is to pin down the ratio of the different types of atoms it contains: this is the compound's **empirical formula**.

This is done using **combustion analysis**. A measured mass of the compound is burned in a mixture of oxygen and helium. Any carbon in the compound is made into carbon dioxide. Any hydrogen in the compound is made into water. Any nitrogen in the compound is initially made into various oxides of nitrogen, but these are then all reduced to nitrogen gas. The masses of carbon dioxide, water and nitrogen gas are measured very precisely to 1 mg (one-thousandth of a gram).

For example, suppose 0.225 g of an unknown organic compound was put through a combustion analysis machine. No nitrogen was produced, so the compound did not contain any nitrogen atoms. 0.706 g of carbon dioxide was made, along with 0.289 g of water.

The mass of one mole of carbon is 44 g.

So 0.706 g of carbon is 0.706/44 moles of carbon = 0.016 moles.

Each mole of water has a mass of 18 g and contains two moles of hydrogen.

0.289 g of water contains 2 × 0.289/18 moles of hydrogen = 0.032 moles.

So the ratio of C:H is 1:2 and the empirical formula is CH_2.

You may be thinking, 'What if there are other atoms, like maybe oxygen?' You can answer this by doing a quick check. 0.016 moles of carbon dioxide has a mass of 0.192 g, and 0.032 moles of hydrogen has a mass of 0.032 g. Added together this makes 0.224 g, which accounts for the entire sample burnt, given that the instrument measures masses to one-thousandth of a gram.

To an experienced chemist, an empirical formula of CH_2 suggests an alkene. That is where the test with bromine water comes in. If this shows the presence of a C=C bond, and the empirical formula is CH_2, then the compound is almost certainly an alkene. The next question is which one.

Mass matters

To work out the **molecular formula** of a substance, you need to know its relative molecular mass as well as the empirical formula. This is usually found using an instrument called a **mass spectrometer**. These are amazing machines that can be used to measure the mass of atoms and molecules.

Mass spectrometers are rather rough on molecules, which sometimes break up in the machine. The relative molecular mass will be the largest result, which will be the whole molecule. Look at the mass spectrometry results for our unknown compound. There are five peaks, the biggest of which is 42. This means that the relative molecular mass of our molecule is 42. The molecule is made up of three 'CH_2' units, so its molecular formula is C_3H_6 and the substance is propene, $CH_3CH=CH_2$. A result!

Mass spectrometers

The mass spectrometer is one of the scientist's most useful analytical tools. It is used to find the relative atomic masses of elements and the relative molecular masses of covalent compounds. Despite the 'spectro' in its name, it does not work by emitting or absorbing light. Instead it 'weighs' atoms or molecules by ionising them and then measuring how easily they are accelerated by an electromagnetic field. Particles with a lower mass are easier to accelerate, while particles with a larger mass are harder to accelerate.

Questions

c Why is a helium and oxygen mixture used, rather than air, in combustion analysis?
d Why must there be an excess of oxygen?

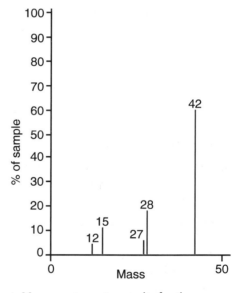

▲ Mass spectrometer results for the unknown alkene.

Key points

- Unsaturated organic compounds, with a C=C group, decolorise bromine water.
- Combustion analysis allows scientists to work out the empirical formula of compounds by accurately measuring the relative amounts of each element in the compound.
- Mass spectrometry is a method for measuring the mass of atoms or molecules. It can be used on elements or compounds.
- You can work out the structure of a molecule from the empirical formula and the relative molecular mass.

Light matters

Matter can react to light in surprising ways. Look at the photo. The 'glow in the dark' bracelet is filled with a compound that takes in or absorbs light at one frequency, and then gives out or emits it at a different frequency. The way that matter interacts with light, and with other electromagnetic radiation like infrared and ultraviolet, is used to analyse matter in a range of modern analytical instruments.

Flame photometry

You have already met flame tests (see section 5.1). A **flame photometer** is a machine that carries out a flame test. The flame photometer in the photo is being used to analyse the amount of Group 1 metals in samples of concrete. The machine can separate the different frequencies of light produced and measure each one. Unlike flame tests, flame photometry is a **quantitative** technique: it can be used to measure the amount of a substance. Flame photometers can use ultraviolet light as well as visible light, because their photosensitive cells can 'see' more wavelengths of light than our eyes can.

▲ 'Glow in the dark' bracelets.

Questions

a What are the scientific words for (i) taking in, and (ii) giving out?
b What is the difference between qualitative and quantitative? (Hint: check section 5.1)
c What is the advantage of using a flame photometer rather than a flame test?
d Suggest a situation when you may use a flame test rather than a flame photometer.

▲ Flame photometer.

Absorption spectroscopy

Flame tests and flame photometers measure what visible or ultraviolet light a substance emits after it is heated up. **Absorption spectroscopy** concentrates on what happens while energy is given to a substance, when the energy is being absorbed.

You have already met one example of this. Helium absorbs some wavelengths of light and not others, so the light from the Sun is missing some wavelengths, as we saw in the introduction to this section. This idea has been developed into a technique used to analyse elements called **atomic absorption spectroscopy**, which can be used both qualitatively and quantitatively.

Atomic absorption spectroscopy is used to measure the amounts of different elements present in different types of steel. Different steels contain different amounts of chromium, manganese and other transition metals. It is these additional atoms that make the steel stainless, or very strong, or unusually hard. Buyers need to know that the steel is exactly right before they build it into tower blocks or cars, for example. Atomic absorption spectroscopy is used to check the composition of the steel. It is also used to measure the amount of different elements in rocks, soils and even foods.

Question

e Why is it important to analyse steel using atomic absorption spectroscopy?

Infrared absorption

Other methods focus on compounds. Some materials absorb infrared (IR) radiation. This time it is the bonds that hold the compound together that are absorbing the energy, rather than the atoms, which absorb visible or ultraviolet light. The energy from the infrared radiation is transferred to the two bonded atoms, making the bond flex or stretch. Different bonds are affected in different ways. This technique is called **infrared spectrometry**.

Look at the infrared spectrometry trace shown in the diagram. If there were no absorption, the line would be horizontal. Each trough shows where a certain type of bond has absorbed, so the IR radiation of that wavelength is missing. For example, the broad trough at 3391 shows the presence of an O–H bond. (You may have noticed that the horizontal axis is 1/wavelength rather than wavelength. This is done so that the important wavelengths are better spread out.)

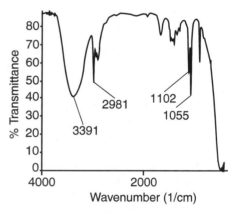

▲ Trace from an infrared spectrometer.

Question

f A scientist has worked out that the molecular formula of an unknown compound is C_2H_6O.
(i) Work out two different possible **structural formulae** for this compound (a structural formula shows how the atoms are grouped together in the molecule, like CH_3CH_3).
(ii) Which of the two possible compounds would have a trough at 3391 if analysed by IR spectrometry?

Ultraviolet absorption

Scientists also check whether ultraviolet (UV) light is absorbed by the sample. In **ultraviolet absorption spectroscopy**, some wavelengths of ultraviolet light are absorbed and not others. Compounds with double bonds like C=O, C=N and C=C absorb UV light. Energy is transferred from the UV light to the shared electrons in the double bond. Ultraviolet light spectrometry is particularly useful for measuring the amount of proteins or DNA in a sample. This is because proteins and DNA contain different combinations of double bonds and absorb UV light of different wavelengths.

Questions

g Why is infrared spectrometry used to analyse compounds rather than elements?
h What is it about proteins and DNA that means they can be analysed using UV spectroscopy?

Key points

- Spectroscopy is the analysis of the light absorbed or emitted by substances. It underlies many of the analytical methods used in modern chemistry.
- Some techniques are used to analyse the elements present, for example atomic absorption spectroscopy.
- Others, such as infrared spectrometry and ultraviolet absorption spectroscopy, tell you about bonds and so are useful when analysing compounds.
- The wavelengths present or absent give scientists qualitative data, while the relative strengths of the signals can give quantitative data.

A very different scene

Chemists from the nineteenth century probably would not recognise a modern chemistry laboratory. Instead of bottles of chemicals and wonderfully shaped glassware, there are machines! These machines have big advantages over the techniques available to a nineteenth-century chemist. For example, they carry out routine procedures perfectly. That frees the scientist to concentrate on more interesting things.

▲ Laboratories have changed!

Technology moves on

The photos show a modern electronic calculator and a mechanical calculating machine from the 1950s. The modern calculator is much smaller and very much faster. It can also carry out much more complicated calculations. Analytical chemists have drawn on these developments and designed machines that incorporate complex electronic equipment. Each year the machines become smaller and quicker. Most modern laboratory instruments have a computer built into them, so they can carry out multi-stage tasks without supervision.

▲ A modern electronic calculator.

Another important development has been a decrease in the amount of sample you need. Many modern instruments can analyse very tiny samples and get reliable results. This is particularly important in forensic science, where the samples are limited to what can be found at the scene of a crime.

Question

a *Suggest ways in which improvements in electronics and computing have made scientists' lives (i) more productive, and (ii) more interesting.*

Equipment in a modern lab

In a modern laboratory there are instruments to carry out routine analysis, many of which have already been discussed in this book. There are titration machines (section 2.4), bomb calorimeters (4.1),

▲ A 1950s mechanical calculator.

mass spectrometers (5.3), machines that carry out combustion analysis (5.3), and even machines that sequence DNA. There are also various types of spectrometers used to analyse materials, including atomic absorption, IR and UV spectrometers (section 5.4). There is also one other group of machines we have not yet discussed: the separation machines, which distil, filter and perform many different types of chromatography.

Question

b Make a table of the machines mentioned in this paragraph and list what they do.

Automated chromatography

We will discuss only one example of a separation machine. It uses **gas–liquid chromatography** to separate substances in very small samples. It is so powerful that 0.000000000001 g of an organic compound can be separated and then identified.

Chromatography always involves two different states of matter, or **phases**. We are familiar with solid, gas and liquid phases of matter. In paper chromatography these are solid and liquid: the paper, and the solvent rising up the paper. In gas–liquid chromatography they are the gas, which is an inert gas like helium, and the liquid, which is inside the coiled column shown in the diagram. The sample is injected into the gas, which carries the sample through the column and then into the detector. Different substances in the sample take different times to pass through the column, so you can identify a substance by how long it takes to come through. You can also tell how much of the substance was present from the size of the peak.

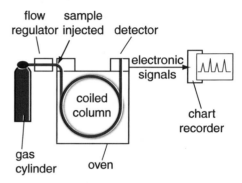

▲ Gas–liquid chromatography.

And finally ...

There is yet another analytical technique that is very important to modern chemists: **nuclear magnetic resonance (NMR) spectroscopy**. This gives scientists information about the hydrogen atoms present in compounds. It can tell the difference between the hydrogen atoms that are present in different groups of atoms, for example in a CH_3 group, a CH_2 group and a CH group, so it is very useful for analysing organic compounds.

NMR spectroscopy uses a very strong magnetic field, produced by a very powerful magnet. The nuclei of hydrogen atoms can act like tiny magnets, interacting with a magnetic field. By analysing these interactions, scientists can learn about the environment of the hydrogen atoms.

Questions

c Why is gas–liquid chromatography particularly useful to forensic scientists? Explain your answer.

d Explain how a scientist could use NMR spectroscopy to tell the difference between CH_3CH_2OH and CH_3OCH_3.

Key points

- Modern labs contain many machines that can carry out analyses, often automatically. The analyses can be done quickly and only very small samples are required to identify substances.
- Developments in electronics and computing have contributed substantially to the automation of spectroscopic and other techniques.
- Gas–liquid chromatography is an example of a technique in which constituents of a mixture are separated physically. It works on very small samples and can give both quantitative and qualitative information about the substances in the mixture.
- NMR spectroscopy is another analytical technique used to analyse compounds.

The challenge

Suppose you are on work experience, working alongside a young chemist called Bernie in a laboratory that specialises in organic chemistry. Yesterday, Bernie showed you all the evidence she had collected about an unknown substance. She wrote notes during your discussion and stuck all the evidence and all her notes on a pinboard. Today she has been called away, but has left you a note asking if you can identify the compound.

Quick wins

Questions

a Look at the photo inset below the topmost graph. It shows the 'before' and 'after' of testing the unknown substance with bromine water. What does this suggest about the compound?

b Look at the mass spectrometry results. What is the relative molecular mass of the compound?

c Look at the infrared spectrometry results. Suggest some groups of atoms that may be present in the compound.

d Look at the NMR spectroscopy results. How many different 'types' of hydrogen atoms are present?

Detailed analysis

Question

e Bernie had started to work through the combustion analysis data. Read what she wrote.

(i) Does Bernie think the compound contains (a) carbon, (b) hydrogen, and (c) nitrogen?

(ii) Why does Bernie think the compound contains some oxygen?

(iii) How many moles of (a) carbon, (b) hydrogen, (c) nitrogen, and (d) oxygen are contained in the sample that was analysed?

(iv) What is the ratio of the different elements present in the unknown compound?

(v) What is the empirical formula of the unknown compound?

(vi) Add up the mass numbers of the atoms in the empirical formula.

(vii) Does this tally with the relative molecular mass from the mass spectrometry?

(viii) Write down the molecular formula of the unknown compound.

Bernie explained yesterday that the way that the molecule broke up during mass spectrometry could give you information about the groups of atoms in the molecule. She scribbled this onto the mass spectrometry results.

Question

f (i) List the groups of atoms Bernie suggested.

(ii) Explain how these relate to the mass spectrometry results.

Put it together

Questions

g Draw out the structures of the possible molecules, showing all the atoms and all the bonds, like H–O–H for water. You may have more than one suggestion. Make sure each one has the correct relative molecular mass.

h Check each of your suggested molecules against the NMR spectroscopy data. Does it have the correct number of different 'types' of hydrogen atoms?

i Write a report for Bernie explaining what you have done. Include a diagram of your final molecule showing all the atoms and the bonds. Give your reasons for the decisions you made.

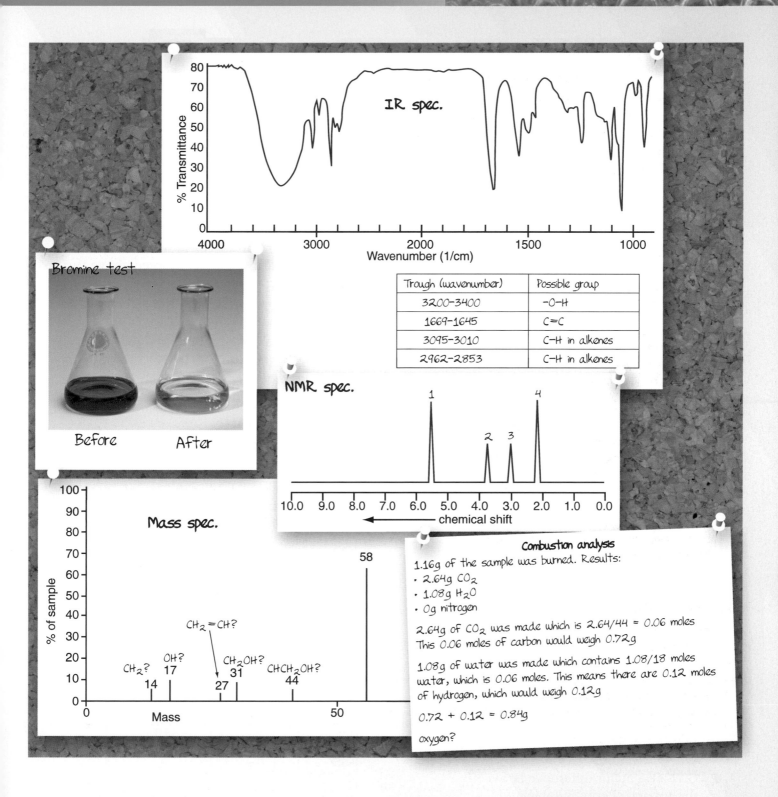

IR spec.

Trough (wavenumber)	Possible group
3200-3400	-O-H
1669-1645	C=C
3095-3010	C-H in alkenes
2962-2853	C-H in alkenes

Bromine test

Before   After

NMR spec.

Mass spec.

CH₂=CH?

CH₂?   OH?   CH₂OH?   CHCH₂OH?
14   17   27   31   44   58

Mass   50

Combustion analysis

1.16g of the sample was burned. Results:
- 2.64g CO₂
- 1.08g H₂O
- 0g nitrogen

2.64g of CO₂ was made which is 2.64/44 = 0.06 moles
This 0.06 moles of carbon would weigh 0.72g

1.08g of water was made which contains 1.08/18 moles water, which is 0.06 moles. This means there are 0.12 moles of hydrogen, which would weigh 0.12g

0.72 + 0.12 = 0.84g

oxygen?

Key points

- Scientists often use many different sources of evidence to solve a problem.
- To identify an unknown organic compound, scientists often use a range of analytical techniques.
- Qualitative and quantitative data come together to decide the structural formula of the compound.

1 Alexandre de Chancourtois was the first scientist to list elements in order of increasing atomic weight. He did this in 1862. The atomic weights he used are shown in the table.

'Element'	Atomic weight	'Element'	Atomic weight
H	1	Si	28
Li	7	Ph	31
Gl	9	S	32
Bo	11	Cl	35
C	12	K	39
Az	14	Ca	40
O	16	Ti	48
NH₄	18	V	51
Na	23	Cr	52
Mg	24	Mn	55
Al	27	Fe	56

a Pick out one 'element' from the table that would not be classed as an element today. Explain your choice. *(2 marks)*

b What are the modern symbols for
i Az, **ii** Bo, **iii** Ph, **iv** Gl? *(4 marks)*

De Chancourtois then looked for patterns in a novel way. He plotted the elements onto cylinders with the atomic weight running down the cylinder. He found that a cylinder that had a circumference of 16, like the one in the diagram, brought similar elements into vertical lines. He had found a repeating pattern every time the atomic weight increases by 16.

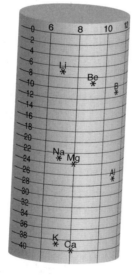

c What **three** modern groups are visible on this side of the cylinder? *(3 marks)*

d Give one major difference between the modern Periodic Table and both de Chancourtois' model and Mendeleev's table. *(1 mark)*

e Suggest one advantage and one disadvantage that a cylinder has compared to a flat periodic table. *(2 marks)*

2 Chlorine is an element in Group 7 of the Periodic Table.

a What is the electronic structure of a chlorine atom? (The Periodic Table in section 1.2 may help you answer this question.) *(2 marks)*

b Chlorine gas is made up of molecules. Draw a 'dot and cross' diagram of the structure of a chlorine molecule, showing only the outer electrons. *(2 marks)*

Chlorine gas was bubbled into a colourless solution of potassium iodide. The solution went brown, then small pieces of solid formed and some purple vapour was observed above the surface of the liquid.

c What evidence is there that a chemical reaction has taken place? *(2 marks)*

d Write **i** a word equation, and **ii** a balanced symbol equation, for the reaction. *(3 marks)*

e Based on this reaction, which is the more reactive element, chlorine or iodine? Explain your answer. *(2 marks)*

f Predict what would happen if chlorine gas was bubbled into a solution of potassium fluoride, and explain your prediction. *(3 marks)*

g When either bromine or iodine reacts, it usually forms halide ions (bromide or iodide). Use your knowledge of atomic structure to suggest why bromine is more reactive than iodine. *(5 marks)*

3 Look at the table showing the hardness and stiffness of various elements.

Element	Moh scale	Young's modulus (MN/cm³)
Ag	2.5	22.4
Au	2.5	41.2
Cr	8.5	15.2
Cs	0.2	4.0
Cu	3.0	19.0
Fe	4.0	16.8
Hg	1.5	–
K	1.4	1.8
Li	0.6	1.1
Na	1.5	2.1
Ni	4.0	19.0
Pt	3.5	45.0
Rb	0.3	3.3
W	7.5	41.1
Zn	2.5	15.2

They are all Group 1 metals or transition metals. The Moh scale is used to judge hardness. It is based on how easily one material scratches another. The hardest material, diamond, is 10 on the Moh scale. Young's modulus is a measure of stiffness, or how easy it is to change the shape of the material when pressure is applied. Low numbers indicate that little pressure is needed.

a Which metal would be the hardest to scratch? *(1 mark)*

b Why is there no Young's modulus for Hg? *(1 mark)*

c Name the Group 1 elements in the table. *(5 marks)*

d Use the data to compare the hardness of Group 1 metals and transition metals. *(3 marks)*

e Use the data to compare the stiffness of Group 1 metals and transition metals. *(3 marks)*

f Use the data and your scientific knowledge and understanding to suggest why manufacturing objects from Group 1 metals is a bad idea. *(3 marks)*

4 Both Group 1 and the transition elements contain metallic elements. As metals, the elements have many properties in common but the differences in their properties have led scientists to place them in two categories rather than one.

a The table shows some properties of metals taken from Group 1 or the transition elements.

Element	A	B	C	D	E	F	G	H
Density (kg/m³)	8920	535	7140	856	28 908	21 090	7874	1532
Melting point (°C)	1084	181	1907	63	1455	1768	1538	39

  i Which **three** elements are most likely to be members of Group 1? *(1 mark)*
  ii Give reasons for your choice of these three elements. *(3 marks)*

b One of these elements has an electronic configuration of 2.8.14.2. Is this a transition element or a Group 1 element? Explain your answer. *(2 marks)*

c One of these elements reacts vigorously when you put a small piece in water, producing a flame. Is this a transition element or a Group 1 element? *(1 mark)*

d One of these elements forms coloured compounds. Is this a transition element or a Group 1 element? *(1 mark)*

e One of these elements is used as a catalyst for many industrial processes. Is this a transition element or a Group 1 element? *(1 mark)*

5 Chemists' understanding of acids was advanced by Arrhenius' ideas about acids and alkalis, published in 1887, and by Brønsted's and Lowry's ideas about acids and bases, published in 1923.

a Write balanced symbol equations to show **i** hydrochloric acid, HCl, acting as an Arrhenius acid, and **ii** sodium hydroxide, NaOH, acting as an Arrhenius alkali. *(2 marks)*

b Write a balanced symbol equation to show the reaction between hydrochloric acid and sodium hydroxide. *(1 mark)*

c Use the Brønsted–Lowry ideas about acids and bases to identify **i** the proton donor, and **ii** the proton acceptor. Mark these on the equation you wrote for question **b**. *(2 marks)*

In 1887 most scientists thought that ions exist in metals only when an electric current is flowing.

d Suggest why scientists found it difficult to accept Arrhenius' ideas about acids and alkalis. *(1 mark)*

By 1923 scientists understood the structure of atoms and ionisation.

e Suggest why scientists quickly accepted Brønsted's and Lowry's ideas. *(1 mark)*

Brønsted and Lowry were not working together. They worked in different research groups in different countries. However, they published very similar ideas at the same time.

f Do you think Brønsted's and Lowry's ideas were as big a step forward as Arrhenius' ideas? Give reasons for your answer. *(2 marks)*

6 The table shows information about three acids.

	Ethanoic acid	Hydrochloric acid	Phosphoric acid
Concentration (mol/dm³)	0.6	0.6	0.6
pH	2.5	0.2	1.2

a Use your understanding of acids to explain why the same concentration of acid does not give you the same pH for the three different acids. *(2 marks)*

b Which is the strongest acid? *(1 mark)*

c Which is the weakest acid? *(1 mark)*

Ethanoic acid is the acid that makes vinegar taste sour. Vinegar contains about 1 mol/dm³ ethanoic acid.

d Use your knowledge of pH to explain why it is safe to put vinegar on your food but not 1 mol/dm³ hydrochloric acid. *(3 marks)*

7 The table shows the values of pH at which three different pH indicators, A (phenolphthalein), B (universal indicator) and C (methyl orange), change colour.

pH	1	2	3	4	5	6	7	8	9	10	11	12	13	14
A														
B														
C														

Which indicator(s) would be suitable for a titration between:

a a strong acid and a weak base?

b a weak acid and a strong base?

c a strong acid and a strong base? *(4 marks)*

8 An environmental scientist was asked to investigate the unexpected death of a rare species of freshwater fish in a lake on an isolated island. She measured the lake water to be pH 5. She took a sample of the water for further analysis but was unable to send it to the mainland due to adverse weather conditions, so she ended up analysing it herself in the local school science laboratory.

a The scientist tested some of the sample with methyl orange and phenolphthalein indicators. The methyl orange indicator stayed yellow and the phenolphthalein indicator stayed colourless. Why? *(2 marks)*

b She decided that she needed to titrate the sample. Which **two** most essential pieces of equipment did she need to find, and why? *(3 marks)*

c She made up some sodium hydroxide solution carefully, so she knew the exact concentration. She dissolved 4.072 g in distilled water and made it up to exactly 1 dm³ with more distilled water. She decided that the concentration of the solution was 0.102 mol/dm³. Show how she came to this answer. (Relative atomic masses: Na = 23, O = 16, H = 1) *(3 marks)*

d She chose phenolphthalein as the indicator for her titration. Why? *(1 mark)*

The scientist titrated 25 cm³ samples of the lake water. These are her results.

Titration	1	2	3
Reading 1	3.2	2.3	0.6
Reading 2	48.7	45.5	43.7
Titre (cm³)	45.5	43.2	43.1

e Which results do you think she should have used to work out her accurate titre? Explain your answer. *(2 marks)*

f Write a balanced symbol equation between H^+ ions and OH^- ions. *(1 mark)*

g What concentration of acid is shown by the titration? *(3 marks)*

Her measurement of pH 5 showed that there were 0.00001 mol/dm³ of H^+ ions in the lake water.

h Explain the discrepancy between the result given by the titration and the pH result. *(2 marks)*

i Was the acid contaminating the lake water a strong acid or a weak acid? Give as much evidence as you can to back up your answer. *(3 marks)*

9 The diagram shows solubility curves for a range of substances.

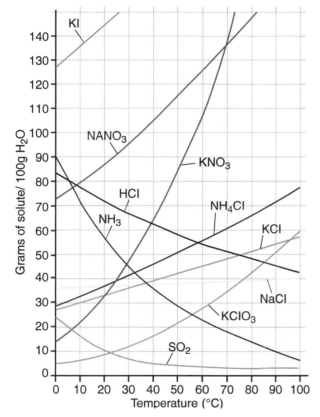

a Which is the most soluble substance at 20 °C?
(1 mark)

b SO₂, HCl and NH₃ are gases. Describe how the solubility of these gases changes with temperature. *(1 mark)*

c The other substances are solids. Describe how the solubility of solids changes with temperature and explain this using your knowledge of particles. *(2 marks)*

d Estimate the mass of potassium nitrate (KNO₃) that would crystallise if you cooled a saturated solution from 70 °C to 10 °C.
(1 mark)

10 The table is about how to process water to make it fit to drink. Match **A**, **B**, **C** and **D** with statements **1–4** in the table. *(4 marks)*

A Chlorination

B Filter bed

C Carbon

D Distillation

1	A method of purifying water.
2	Removes small pieces of suspended solid from the water.
3	Sterilises the water by killing microorganisms.
4	A filter that improves the taste of the water.

11 Activated carbon filters can be used to remove impurities from water. The table shows information about an activated carbon filter.

Impurity	Removes all	Removes some	Removes little or none
Arsenic			✓
Bacteria and viruses			✓
Chlorine	✓		
Fluoride			✓
Nitrates			✓
Radon	✓		
Sediments		✓	
Iron			✓
Volatile organic compounds	✓		

a Study the table. Would such a carbon filter be enough to treat water from the environment? Explain your answer. *(2 marks)*

b Suggest why a carbon filter improves the taste of tap water. *(1 mark)*

c Would you recommend this water filter for the following potential buyers? Give your reasons in each case.

i A family living in a house built on granite, where radon gas had been detected in the air. *(2 marks)*

ii A person living in an area where fluoride was added to the water, but who believed that fluoride was unhealthy. *(2 marks)*

iii A fish enthusiast whose tropical fish were very sensitive to organic compounds like hydrocarbons. *(2 marks)*

iv A mining company that mines heavy metals and wants to use a local groundwater for drinking water. *(2 marks)*

12 George's parents are considering buying a water softener from Soft Water plc. The representative from the company has tested the water and said that it is hard. George decides to carry out his own investigation to see whether the water in his home is hard. He does an Internet search and finds these definitions of hardness.

	Conc. of calcium carbonate (mg/dm³)			
	0–74	75–149	150–299	300+
Hardness rating	soft	medium-hard	hard	very hard

George decides to find the concentration of the calcium and magnesium ions by titration. He finds out that a substance called EDTA reacts with calcium and magnesium ions. Although EDTA has a complex formula, the reaction can be represented like this, where X^{2+} is a calcium or magnesium ion:

$$EDTA + X^{2+} \rightarrow EDTA\,X^{2+}$$

He finds out that 'calmagite' is a suitable indicator. It is red when calcium or magnesium ions are present, and blue when they have all reacted. The instructions with the indicator say that the end-point is when the indicator is blue with no hint of red.

George collects his samples of water from the kitchen tap, the bathroom tap, and the filter jug that his Mum keeps in the fridge. He carries out some quick tests to find a suitable concentration of EDTA solution:

Concentration of EDTA (mol/dm³)	0.0001	0.001	0.01
Vol. of EDTA to react with 10 cm³ water	87	9	1

He decides to use 0.001 mol/dm³ EDTA solution, which he places in a burette. He measures out 25 cm³ of water into a conical flask using a pipette. He adds three drops of indicator and puts the conical flask on a white tile. Here are his titration results:

Source of sample	Titration number and titre (volume of EDTA solution) (cm³)				
	Rough	1	2	3	Average
Kitchen	28.2	24.7	24.4	24.5	24.45
Bathroom	24.9	23.8	23.7		23.75
Filter jug	25.4	24.7	24.6		24.65
100 mg/dm³ CaCO₃	26.7	25.1	25.2		25.15

George changed one variable (the independent variable) to find out how this altered another variable (the dependent variable).

a In George's investigation, what is the independent variable? *(1 mark)*

b What kind of variable was this? *(1 mark)*
  A a continuous variable
  B a discrete variable
  C an ordered variable
  D a ranked variable

c List **two** variables that George kept the same in his investigation. *(2 marks)*

d George included a control in his investigation.
  i What was the control? *(1 mark)*
  ii Why did he include it? *(1 mark)*

e Explain why:
  i George decided to use 0.001 mol/dm³ EDTA solution. *(1 mark)*
  ii He used 25 cm³ of water for his titration rather than 10 cm³. *(1 mark)*
  iii He placed a white tile under the conical flask. *(1 mark)*

George analysed his results for the 100 mg/dm³ calcium carbonate solution:
- 1 cm³ of 0.001 mol/dm³ EDTA contains 0.000001 moles of EDTA.
- 25.15 cm³ contains 0.00002515 moles of EDTA.
- So 25 cm³ of kitchen tap water contains 0.00002515 moles of $CaCO_3$.
- So 1000 cm³ of kitchen tap water contains 0.001006 moles of $CaCO_3$.
- 1 mole of calcium carbonate has a mass of 100 g.
- So 0.001006 moles of $CaCO_3$ has a mass of 0.1006 g or 100.6 mg.

f i What concentration of calcium carbonate solution was George using? *(1 mark)*
  ii What concentration for the calcium carbonate solution did George work out from the titration? *(1 mark)*
  iii What does this tell you about the accuracy of George's method? *(2 marks)*

g If a titre of 25.15 cm³ indicates a hardness of 100 mg/dm³, what is the hardness of:
  i the kitchen water? *(1 mark)*
  ii the bathroom water? *(1 mark)*
  iii the filtered water? *(1 mark)*

h Can you think of any improvements to George's investigation to make his results more reliable? *(1 mark)*

i In your opinion, does George's family need a water softener? Back up your decision using George's results and the data given above.

(2 marks)

13 A chemical engineer is developing an industrial process during which $1000\,dm^3$ of nitric acid solution is neutralised by $1000\,dm^3$ of sodium hydroxide solution. He remembers that the reaction is exothermic. He decides to carry out a small-scale trial.

a Why is a small-scale trial a good idea? *(1 mark)*

He mixes $1\,dm^3$ nitric acid solution with $1\,dm^3$ sodium hydroxide solution. The temperature rise is 12.4°C.

b What is the total volume? *(1 mark)*

c Given that it takes 4.2 J to raise $1\,cm^3$ water by 1°C, how much energy has been added to the mixture? *(2 marks)*

d The engineer uses standard data to find that the expected enthalpy change is –15.8 kJ/mol. Suggest a reason for this discrepancy. *(1 mark)*

e If the energy added to the mixture had been –115.8 kJ, what temperature rise would have occurred? *(2 marks)*

f What temperature rise should the engineer expect when the process is scaled up?
   A less than 12.4°C
   B 12.4°C
   C between 12.4°C and 13.8°C
   D 12400°C *(1 mark)*

14 The diagram shows self-heating soup in a can.

1 Water
2 Lime
3 Standard size tin of soup
4 Clamp to hermetically seal lime and water compartments
5 Easy opening system
6 Heat insulator
7 Waterproof separator

The heating process is started by piercing holes from the rim of the can through the waterproof separator between the water and the lime. Calcium oxide (lime) reacts with water to produce calcium hydroxide in an exothermic reaction. Enough thermal energy needs to be transferred to the soup to make it boil. A standard tin contains about 450 g. The energy released has to heat the water in the soup and the water that is reacting, a total of 500 g.

Scientists developing this product investigated the reaction using a calorimeter. Their results are shown in the table.

Mass of calcium oxide (g)	Total mass of water heated (g)	Starting temp. (°C)	Max. temp. (°C)	Temp. rise (°C)
50	500	5	31	26
100	500	5	47	42
150	500	5	75	70
200	500	5	78	73

a In this investigation, what were:
   i the independent variable?
   ii the dependent variable? *(2 marks)*

b Give **two** variables that were kept constant. *(2 marks)*

c What type of variable was the independent variable? Choose from: *(1 mark)*
   A continuous
   B discrete
   C ranked
   D ordered

d If 50 g of calcium oxide reacted to produce a temperature rise of 26 °C, predict the temperature rise you would expect from:
   i 150 g
   ii 200 g. *(2 marks)*

e Suggest a reason why the temperature rise for 200 g of calcium oxide is much lower than you expected. *(1 mark)*

The scientists decide to use 200 g of calcium oxide and 100 g of water for the reaction. The total mass of water heated is now 550 g. The water is heated from 5 °C to boiling point.

f If it takes 4.2 J to heat 1 g of water by 1 °C, calculate the energy transferred to the water when it is heated to 100 °C. *(4 marks)*

15 Manufacturers are working on cars powered by hydrogen. Hydrogen reacts with oxygen in an exothermic reaction:

$$2H_2 + O_2 \rightarrow 2H_2O$$

a Suggest why burning hydrogen might cause fewer pollution problems than burning fossil fuels. *(4 marks)*

Hydrogen can be made by splitting water molecules. The water-splitting reaction can be represented by the following equation:

$$2\ H-O-H \rightarrow 2\ H-H\ +\ O=O$$

The table gives some bond energies:

Bond	Bond energy (kJ per mole)
O–H	464
H–H	426
O=O	498

b Use the bond energies to estimate the energy transferred in this reaction. *(3 marks)*

c Explain, in terms of bond energies, why this reaction is endothermic. *(2 marks)*

At the moment, the following reaction is used to make most of the hydrogen used for fuel. Methane found in natural gas is the raw material.

$$CH_4 + 2H_2O \rightarrow 4H_2 + CO_2$$

Certain species of green algae produce hydrogen in the presence of sunlight. In the summer of 2001, researchers changed the photosynthetic process of spinach plants to produce hydrogen.

d Give **two** reasons why producing hydrogen by photosynthesis would be better than producing hydrogen from natural gas. *(2 marks)*

16 Use the information in the table to predict whether each of these reactions is endothermic or exothermic and to estimate the enthalpy change for each reaction.

Bond	Average bond energy (kJ/mol)
C–H	413
O=O	498
C=O	805
H–O	464
C–C	347
C≡C	838
O–O	138

Structures of molecules	
H_2O	H–O–H
H_2	H–H
O_2	O=O
CO_2	O=C=O
H_2O_2	H–O–O–H
C_2H_2	H–C≡C–H

a $2H_2O \rightarrow 2H_2 + O_2$ *(4 marks)*
b $2H_2O_2 \rightarrow 2H_2O + O_2$ *(4 marks)*
c $2C_2H_2 + 5O_2 \rightarrow 4CO_2 + 2H_2O$ *(4 marks)*

17 Four labels have come off four bottles:

Aluminium sulfate solution $Al_2(SO_4)_3$(aq)	Magnesium chloride solution $MgCl_2$(aq)
Magnesium sulfate solution $MgSO_4$(aq)	Ammonium chloride solution NH_4Cl(aq)

Describe and give the results of the **chemical** tests that you would do to identify which bottle contains which substance. *(6 marks)*

18 0.88 g of a hydrocarbon were analysed by combustion analysis. 2.64 g of carbon dioxide and 1.44 g of water were made.

(Relative atomic masses: C = 14, H = 1, O = 16, N = 14)

a How many moles of carbon dioxide were made? *(3 marks)*
b How many moles of carbon atoms were there in the sample? *(1 mark)*
c How many moles of water were made? *(3 marks)*
d How many moles of hydrogen atoms were there in the sample? *(1 mark)*
e How do you know that the sample contains only carbon and hydrogen? *(1 mark)*
f What is the empirical formula of the substance analysed? *(2 marks)*

A mass spectrometer is used to show that the relative molecular mass of the substance is 44 g/mole.

g What is the molecular formula of the substance? *(1 mark)*

19 The table is about techniques used to analyse substances. Match **A**, **B**, **C** and **D** with statements **1–4** in the table. *(4 marks)*

A infrared spectrometry
B mass spectrometry
C atomic absorption spectroscopy
D NMR spectroscopy

1	Gives information about hydrogen nuclei and where they are in a molecule.
2	Gives information about bonds in molecules.
3	Gives the mass of elements and compounds.
4	Gives information about atoms based on the wavelengths they absorb.

20 Mass spectrometry is used to analyse the atoms present in an element. The first diagram shows the mass spectrometry trace of the element zirconium.

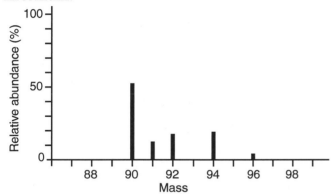

a How many different isotopes are present?

(1 mark)

b Which is the most abundant isotope?

(1 mark)

c Estimate the relative atomic mass of zirconium. Choose between:

A 89–90   B 90–91
C 91–92   D 92–93   *(1 mark)*

The second diagram shows mass spectrometry of bromine molecules. Some of the molecules have broken up.

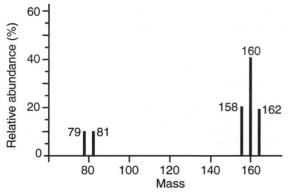

d Explain why there are three peaks, at 158, 160 and 162, that show the bromine molecules present. *(3 marks)*

e Out of 100 atoms of chlorine, 76 have a mass number of 35, and 24 have a mass number of 37. Show how you could use this information to estimate the relative atomic mass of chlorine.

(3 marks)

21 The fire brigade has been called to a derelict factory that is due for demolition. A large cupboard has been found that contains glass bottles of unknown substances. Most of the labels have come away because of the damp conditions, but there is a separate small cupboard with a label 'poisons'. The fire brigade calls in a specialist team of scientists to dispose of the chemicals safely.

The scientists take an inventory of the bottles, giving each a code number. The table shows the first 12 bottles, which were not from the poisons cupboard.

Code no.	Description
1	White crystalline solid
2	White crystalline solid
3	Blue crystalline solid
4	Silvery cubes under oil
5	White crystalline solid
6	Grey crystals with purple vapour
7	White powder
8	White crystalline solid
9	Brown crystalline solid
10	White crystalline solid
11	Silvery cubes under oil
12	White crystalline solid

Suggest, with reasons, which bottles may contain the following substances and the tests you would carry out to confirm whether your suggestion is correct.

a Iodine *(2 marks)*
b Copper II sulfate *(2 marks)*
c Potassium *(3 marks)*
d Aluminium chloride *(4 marks)*
e Calcium carbonate *(4 marks)*

Glossary

absorbed Taken in, for example when light energy is taken in by a substance.

absorption spectroscopy The production, measurement and analysis of electromagnetic spectra produced as a result of the absorption of energy by substances.

acid rain Rain that has a pH of less than 5.6 because of dissolved pollutants, usually sulfur dioxide.

activation energy The minimum energy needed before a chemical reaction can occur.

addition polymerisation Polymerisation that occurs when alkene monomers are made to combine by opening their double bond; no other chemicals are involved in the main reaction.

alkali A substance that dissolves in water to give a solution of hydroxide ions with a pH of over 7.

alkali metals Group 1 elements.

alkane A hydrocarbon which only has single C–C or C–H bonds, e.g. ethane C_2H_6.

alkene A hydrocarbon with one (or more) C=C double bonds, e.g. ethene C_2H_4.

alloy Carefully blended metal mixture with specific properties.

atom The smallest part of an element that still has the properties of that element.

atom economy How much useful product is produced in a reaction compared to the total mass of atoms used.

atomic absorption spectroscopy The production, measurement and analysis of electromagnetic spectra produced as a result of the absorption of energy by atoms.

atomic number (Z) The number of protons in an atom; same as the proton number.

atomic weight An out-of-date way of giving the mass of an atom relative to the mass of hydrogen. Now replaced by relative atomic mass.

base A substance that reacts with an acid, reducing its acidity.

biodegradable Something that can be broken down easily by natural, biological processes.

biodiesel A renewable fuel made from vegetable oil that can be used in place of diesel.

bomb calorimeter Apparatus for measuring the thermal energy released when a substance is burned.

bond energy The average energy associated with each type of covalent bond, e.g. C–C, obtained by experiment.

brine A solution of sodium chloride in water, such as sea water.

burette An instrument for measuring and dispensing a variable volume of liquid.

calibrated For a measuring instrument, having a scale, e.g. the degrees on a thermometer.

calorimeter Apparatus for measuring the thermal energy released or taken in during a change.

causal An occurrence or variable that results in a change to another variable.

cement A powder which sets hard when mixed with water. It is made by heating limestone with clay.

chemical bonds Bonds that form between atoms during chemical reactions that hold molecules and other compounds together.

chromatography A way to separate different dyes.

closed system A reacting system is closed if none of the products can escape.

combustion analysis A standardised method of burning an organic substance and analysing the products to find out the amount of each element present in a known mass of the substance.

compounds Atoms of two or more different elements chemically joined together to form a new substance with new properties.

concrete A building material made by mixing cement, sand and gravel with water; it sets to become a very hard 'artificial rock'.

continental drift The exceedingly slow movement of the continents across the surface of the Earth.

convection Transfer of thermal energy in a gas or liquid by the movement of particles from place to place. Hotter, less dense regions float and cooler, denser regions sink.

covalent bond The attraction between atoms that are sharing electrons within a molecule.

covalent compound A compound formed when atoms from two or more elements are joined by covalent bonds.

cracking The process by which long-chain hydrocarbons are broken up into shorter and more useful hydrocarbons.

crust The thin, hard and brittle outer layer of the Earth.

crystal A solid formed when particles stack up in a regular way.

displacement A type of chemical reaction in which one element reacts with a compound of a second element and the result is the second element and a compound of the first element.

distillation Separation of a mixture of liquids with different boiling points.

double decomposition A chemical process whereby two soluble metal salts 'swap partners' in solution to form an insoluble compound that precipitates out of solution.

electrolysis The tearing apart of a molten (or dissolved) ionic compound using electricity.

electron shells The positions that electrons can occupy around an atom – same as energy levels.

electrons Particles much smaller than an atom, with a tiny mass and a single negative charge. Electrons move through metals when a current flows.

element Something made of one type of atom only.

emitted Given out, for example when light energy is given out by a substance.

empirical formula A combination of symbols showing the ratio of the different types of atoms in a compound.

emulsifier A chemical that helps to stop an emulsion from separating.

emulsion A mixture of tiny droplets of oil in water (or water in oil).

endothermic A reaction that takes in energy.

end-point The point in a titration when the reactants are exactly balanced so that they will completely react with each other with no excess of either. Often marked by the change of colour of an indicator.

energy levels The positions that electrons can occupy around an atom – same as electron shells.

energy value The energy released per gram of fuel or food burned.

enthalpy, *H* A measure of the energy of a substance.

equilibrium The balance point in a reversible reaction reached when the rates of the forward and back reactions are equal.

ethanol The 'alcohol' in alcoholic drinks made by fermenting sugar. It may also be used as a fuel.

exothermic A reaction that gives out energy.

filter beds Horizontal layers of porous material through which water is passed to remove tiny, suspended pieces of solid.

flame photometer A machine that carries out flame tests and uses the light emitted to measure the amount of metal ion present.

flame test A test in which a small amount of a salt is put in a flame; the colour observed is typical of the metal ion present.

formula mass (M_r) The sum of the relative atomic masses of all the atoms in a compound (e.g. for water H_2O: $2 \times 1 + 16 = 18$).

fractional distillation A form of distillation where only a partial separation of the liquids in a mixture is obtained.

fractions The different liquids produced from a complex mixture such as crude oil by fractional distillation.

fuel cell A device that converts the energy released when a fuel is oxidised directly and continuously into electrical energy.

gas–liquid chromatography The separation of a mixture into its component substances based on their relative solubility in a gas and a liquid.

gel A material with a grid-like structure that can trap water.

global dimming Sunlight reaching the Earth is weakened due to atmospheric pollution.

global warming Increased proportions of greenhouse gases such as carbon dioxide are causing the average temperature at the surface of the Earth to rise.

greenhouse effect Gases such as carbon dioxide help to trap heat energy in the atmosphere.

greenhouse gas Gases such as carbon dioxide that help to trap heat energy in the atmosphere.

half equation The reaction at one electrode in an electrolytic cell.

halide An ionic compound of one of the halogens, such as chlorine.

halogen A member of Group 7 of the Periodic Table, such as chlorine.

hard Water that contains high levels of dissolved calcium and magnesium salts is described as 'hard'.

hydrazine A substance, N_2H_4, that is used as a fuel in rockets.

hydrocarbon Compounds made of carbon and hydrogen atoms only.

hydrogels Strong gels that can be used for contact lenses.

hydrogen bonds Particularly strong intermolecular forces (bonds) that can occur between molecules that contain an H–O, H–N or H–F bond. In ice, one water molecule hydrogen-bonds to four other water molecules.

hydrogenation A chemical reaction where a hydrogen molecule joins with another compound, e.g. the hydrogenation of unsaturated oil to make saturated fat.

hydrolysis A chemical reaction where a water molecule joins with another compound, e.g. the hydrolysis of ethene to make ethanol.

infrared spectrometry The production, measurement and analysis of infrared (IR) spectra produced as a result of the absorption of energy by bonds.

inorganic A substance that is not organic.

intermolecular force The weak force of attraction between uncharged molecules.

ion Charged particles.

ion exchange A process by which one ion in a solution is exchanged for another.

ion exchange resin An insoluble support with a charged surface, on which one ion is exchanged for another of the same charge.

ionic bond The attraction between oppositely charged ions in a compound.

ionic compound A compound formed by the electrostatic attraction between oppositely charged ions.

ionise For an atom, to gain or lose electron(s) and become charged (an ion).

isotope Atoms of an element come in different forms or isotopes, depending on the numbers of neutrons they have in their nuclei.

lattice The regular arrangement of particles, e.g. in a crystal.

limescale An insoluble deposit of calcium and magnesium salts that occurs on surfaces in hard water areas.

litmus A pH indicator originally isolated from lichen, which is pink in acidic solutions and blue in alkaline solutions.

macromolecule Giant structures where all the bonds are strong covalent bonds, such as diamond.

mantle The middle, soft, rocky layer of the Earth which can move very slowly.

mass number (A) The number of protons added to the number of neutrons in an atom.

mass spectrometer An instrument used to find the relative mass of atoms, ions and molecules.

meniscus The curved surface of a liquid in a tube.

mole The amount of a substance containing 6.02×10^{23} particles (i.e. the relative atomic mass or formula mass in grams).

molecular formula A combination of symbols showing the different types of atoms and the number of each type of atom in a molecule of a covalent substance.

molecules Particles made from atoms joined by covalent bonds.

monomer The small molecules that are joined up to form a polymer.

monopropellant A rocket fuel that reacts on its own, without the need for another reactant.

mortar A paste made from slaked lime and water that was once used to stick bricks together.

nanometre 1 billionth of a metre (1 millionth of a millimetre).

nanoparticles Very small particles just a few tens of nanometres across or less.

nanotechnology Technology based around the special properties of nanoparticles.

neutralisation A chemical reaction between an acid and alkali (or base) that produces a neutral salt.

neutron A sub-atomic particle with no electric charge and a relative mass of 1.

noble gas A gas from Group 0 of the Periodic Table, e.g. helium, argon, neon.

nuclear magnetic resonance (NMR) spectroscopy The analysis of how the nuclei of certain atoms (usually hydrogen) interact with a strong magnetic field. Different nuclei of the same type of atom interact differently according to their environment, e.g. the surrounding atoms in the molecule.

nucleus The central part of the atom containing the proton(s) and, for all except hydrogen, the neutrons; it contains most of the mass of the atom.

ohm (Ω) The unit of electrical resistance.

open system A reacting system is open if some of the products can escape (or are removed).

ore A natural compound of a metal from which the metal can be extracted.

organic A covalent compound containing carbon and hydrogen.

oxidation A type of chemical reaction. When a substance is oxidised it gains oxygen, loses hydrogen or loses electrons.

oxidation Reaction with oxygen, e.g. iron is oxidised when iron oxide forms during rusting. Oxidation may also be defined as the loss of electrons.

oxonium ion An H_3O^+ ion, which is a combination of a water molecule and a hydrogen ion (a proton).

percentage yield The actual yield as a percentage of the theoretical yield for a reaction.

periodic table Table of all the elements arranged to show patterns in their properties. The modern Periodic Table is arranged in order of increasing atomic number and reflects the structure of the atoms.

periodicity Pattern that repeats at the same intervals.

pH scale A number scale to show the strength of acids or alkalis. 1 is the strongest acid, 7 is neutral, 14 is the strongest alkali.

phase Distinct state of matter, e.g. gas, liquid, solid.

photosynthesis A series of reactions in which plants use light energy to make sugars.

pipette An instrument for measuring and dispensing volumes of liquid. Many pipettes are made to measure and dispense a single volume, but some are graduated and can dispense a variety of volumes.

plate tectonics Theory used to explain continental drift.

polymerisation The chemical process that joins monomers together to make a polymer.

polymers Very long-chained molecules made by joining many small molecules together.

precipitate An insoluble solid that sometimes forms when two reacting solutions are mixed.

properties What a material is like, e.g. melting point, hardness, strength, etc.

proton A sub-atomic particle with a positive charge and a relative mass of 1.

proton acceptor A substance that accepts a hydrogen ion (a proton) from another substance.

proton donor A substance that gives a hydrogen ion (a proton) to another substance.

proton number Another name for the atomic number.

qualitative Identifying, but not measuring, characteristics.

qualitative analysis In chemistry, methods that tell you what substances are present, but not the amount of each substance.

quantitative Measuring the amount of a characteristic or characteristics.

quicklime Calcium oxide (CaO), a strong alkali formed by heating limestone.

rate of reaction How much reactant is used up (or product formed) divided by the time taken.

recommended dietary allowance (RDA) The amount of energy and nutrients recommended per person per day.

recycle To re-use materials over and over again.

reduction A type of chemical reaction. When a substance is reduced it loses oxygen, gains hydrogen or gains electrons, e.g. iron oxide is reduced to iron.

relative atomic mass (A_r) The mass of an atom compared to 1/12th of the mass of a carbon-12 atom.

rock salt The natural mineral form of sodium chloride.

salts Ionic compounds formed from positive ions (usually metallic) and negative ions from an acid.

saturated A carbon-chain molecule that only has single bonds between the carbon atoms.

slaked lime Calcium hydroxide ($Ca(OH)_2$), made by adding water to quicklime.

smart material A material that can change its properties if its environment changes.

soft Water that does not contain high levels of dissolved calcium or magnesium ions is described as 'soft'.

solar spectrum The light given out by the Sun, spread out using prisms.

solubility The mass of a substance that can dissolve in 100 g of solvent.

solubility curve A graph of solubility against temperature.

spectroscopy The production, measurement and analysis of electromagnetic spectra produced as a result of the emission or absorption of energy by substances.

stable A substance that is unlikely to react is described as stable. Stable is the opposite of reactive.

state symbols These symbols are often added to symbol chemical equations: (s) solid, (l) liquid, (g) gas, (aq) solution in water (aqueous solution).

still Apparatus used for distillation.

strong acid An acid that fully ionises in water, producing the maximum number of hydrogen ions and a low pH.

structural formula The molecular formula written to show how the atoms are grouped in the molecule, e.g. CH_3CH_2OH.

sublimes When a solid turns straight into a gas (or vice versa).

sustainable development Development that conserves natural resources and so can be used successfully into the future.

tectonic plates Massive sections of the Earth's crust that gradually move around the Earth's surface, transporting the continents.

theoretical yield The theoretical amount of a product you should get from a reaction based on its symbol equation.

thermal decomposition Breaking down a chemical compound by heating.

titration A method for measuring concentrations of substances in solution.

titre The volume of the reactant of known concentration used in a titration.

transition elements The block of metallic elements that lie between Group 2 and Group 3 in the modern Periodic Table.

transition metals The 'everyday' metals such as iron and copper, found in the central block of the Periodic Table.

tsunami A large wave generated by an undersea earthquake.

ultraviolet absorption spectroscopy The production, measurement and analysis of ultraviolet (UV) spectra produced as a result of the absorption of energy by double bonds.

universal indicator The most commonly used pH indicator, made up of a mixture of three indicators that change colours at different pHs. It is green at neutral pH (pH 7).

unsaturated Organic compounds that contain double bonds, e.g. C=C, are unsaturated.

weak acid An acid that does not fully ionise in water, producing less than the maximum number of hydrogen ions and a relatively high pH (although still below 7).

yield The amount of product you get; often expressed as a percentage of what is theoretically possible.

ΔH Change in enthalpy: the thermal energy taken in ($+\Delta H$) or given out ($-\Delta H$) during a change.

Index